FOR THE IB DIPLOMA

Environmental Systems and Societies
Study and Revision Guide

SECOND EDITION

Andrew Davis
Garrett Nagle

Dedications

For my family, and with thanks to Danny, Cecilia and Yvonne Chew, Dr Arthur Chung, Dr Chey Vun Khen, and to all my friends and colleagues in Sabah. This book is dedicated to the memory of Dr Clive Marsh.

A. J. Davis

With thanks to Angela, Rosie, Patrick and Bethany for their continued support, patience and good humour.

G. E. Nagle

Our thanks to So-Shan Au for her help and guidance throughout this project.

The Publishers would like to thank the following for permission to reproduce copyright material.

Photo credits

p.1 © Alfred Eisenstaedt/The LIFE Picture Collection/Getty Images; **p.2** © Martin Bond/Alamy Stock Photo; **p.7** © Stéphane Bidouze – Fotolia; **p.10** © Ingram Publishing Limited; **p.13** © Christinezenino/http://creativecommons.org/licenses/by/2.0/http://www.flickr.com/photos/chrissy575/3977247173_130424; **p.20** © So-Shan Au; **p.23** *t* © Toa55/Getty Images/iStockphoto/Thinkstock, *b* © Danicek – Fotolia; **p.28** © EcoView – Fotolia; **p.30** *t* © Dante Fenolio/Science Photo Library, *l* © Kaz Chiba/Photodisc/Getty Images, *r* © Andrew Davis; **p.32** © Andrew Davis; **p.33** © inka schmidt – Fotolia; **p.54** © Stockbyte/ Photolibrary Group Ltd; **p.55** © Ansud – Fotolia; **p.57** © Andrew Davis; **p.59** © Andrew Davis; **p.62** © Andrew Davis; **p.63** NASA's Earth Observatory/http://earthobservatory.nasa.gov/Features/WorldOfChange/deforestation.php; **p.65** © Villiers – Fotolia; **p.66** © Professor T. Keith Philips, Ph.D. http://zookeys.pensoft.net/lib/ajax_srv/article_elements_srv.php?action=zoom_figure&instance_id=11&article_id=6183; **p.68** © So-Shan Au; **p.76** © Andrew Davis; **p.77** © chameleonseye/Thinkstock/iStock/Getty Images; **p.78** © Andrew Davis; **p.80** © Andrew Davis; **p.86** © Andrew Davis; **p.87** © FLPA/Alamy Stock Photo; **p.89** ©Getty Images/iStockphoto/Thinkstock; **p.100** © Garrett Nagle; **p.103** © Garrett Nagle; **p.120** © Garrett Nagle; **p.138** © Stockbyte/ Photolibrary Group Ltd; **p.142** © Garrett Nagle; **p.157** Scuplture © Paul Bonomini, photo © Garrett Nagle.

Orders: please contact Hachette UK Distribution, Hely Hutchinson Centre, Milton Road, Didcot, Oxfordshire, OX11 7HH. Telephone: +44 (0)1235 827827. Email education@hachette.co.uk Lines are open from 9 a.m. to 5 p.m., Monday to Friday. You can also order through our website: www.hoddereducation.com

ISBN: 9781471899737

© Andrew Davis and Garrett Nagle 2017

First published in 2013

This edition published 2017 by

Hodder Education,

An Hachette UK Company

Carmelite House

50 Victoria Embankment

London EC4Y 0DZ

www.hoddereducation.com

Impression number   10 9 8

Year   2022

Cover photo © SasinTipchai/Shutterstock

Illustrations by Integra Software services and Aptara Inc.

Typeset in GoudyStd 10/12 pts. by Aptara Inc.

Printed in India.

A catalogue record for this title is available from the British Library.

Contents

How to use this revision guide

Welcome to the *Environmental Systems and Societies for the IB Diploma Study and Revision Guide*. This book will help you plan your revision and to work through it in a methodological way. It follows the Environmental Systems and Societies (ESS) syllabus topic by topic, with revision and practice questions at the end of each section to help you check your understanding.

■ Features to help you succeed

SIGNIFICANT IDEAS

There are eight topics in the IB ESS syllabus, each divided into several sub-topics. Each sub-topic begins with a list of 'Significant ideas', which are overarching principles that summarise the content of each section.

Expert tip

These tips give advice that will help you boost your final grade.

Worked example

Some parts of the course require you to carry out mathematical calculations, plot graphs, and so on. These examples show you how.

Common mistake

These identify typical mistakes that students make and explain how you can avoid them.

CASE STUDY

Parts of the syllabus require you to use case studies. Examples are given in the relevant sections of the book.

■ QUICK CHECK QUESTIONS

Use these questions to make sure that you have understood a topic. They are short, knowledge-based questions that use information directly from the text.

Keyword definitions

Definitions are provided on the pages where the essential key terms appear. These key words are those that you can be expected to define in exams. A **glossary** of these essential terms, highlighted throughout the text, is given at the end of the book.

EXAM PRACTICE

Practice exam questions are provided at the end of each section. Use them to support your revision and practise your exam skills.

You can keep track of your revision by ticking off each sub-topic heading in the book. There is also a checklist at the end of the book. Use this checklist to record progress as you revise. Tick each box when you have:

- revised and understood a topic
- tested yourself using the **Quick check questions**
- used the **Exam practice** questions and gone online to check your answers.

Use this book as the cornerstone of your revision. Don't hesitate to write in it and personalise your notes. Use a highlighter to identify areas that need further work. You may find it helpful to add your own notes as you work through each topic.

Getting to know Paper 1 and Paper 2

At the end of your ESS course you will sit two papers. Paper 1 is worth 25% of the final marks and Paper 2, 50% of the final marks. The other assessed part of the course (25%) is made up of the Internal Assessment (practical work), which is marked by your teacher.

There are no options in ESS and therefore all topics need to be thoroughly revised for both papers. Here is some general advice for the exams:

- Make sure you have learned the command terms (e.g. evaluate, explain, outline). There is a tendency to focus on the content in a question rather than the command term, but if you do not address what the command term is asking of you then you will not be awarded marks. (Command terms are covered on pages vii–xi.)

- Answer all questions and do not leave gaps.
- Do not write outside the answer boxes provided – if you do so this work will not be marked.
- If you run out of room on the page, use continuation sheets and indicate clearly that you have done this on the cover sheet. (The fact that the question continues on another sheet of paper needs to be clearly indicated in the text box provided.)
- Plan your time carefully before the exams – this is especially important for Paper 2 (see below).

■ Paper 1

Paper 1 (1 hour) contains short-answer and data-based questions relating to a previously unseen case study. The total number of marks for this paper is 35. The case study is contained in a resource booklet, in which you will be given a range of data in various forms (e.g. maps, photos, diagrams, graphs and tables). Questions will test your knowledge of the syllabus and your ability to apply this to the new case study. You are required to answer a series of questions, which can involve a variety of command terms, by analysing these data. Questions will be based on the analysis and evaluation of the data in the case study, and all of the questions are compulsory. The questions test assessment objectives 1, 2 and 3 (for information on assessment objectives, see pages vii–xi). Remember:

- The size of the boxes provided gives an indication of the length of answer expected – make sure your answers are concise.
- Look carefully at the number of marks awarded for each question. For example, if 2 marks are awarded the examiner is looking for two different points.

■ Paper 2

Paper 2 (2 hours) is in two sections: Section A (25 marks) and Section B (40 marks). Total marks for the paper are therefore 65. The questions test assessment objectives 1, 2 and 3. Section A contains short-answer and data-based questions; because this section can cover any aspect of the course, it is essential that you thoroughly revise the whole syllabus so that you can tackle any questions that come up. In Section B you must answer two structured essay questions from a choice of four. Each essay is worth 20 marks. The final part of each essay in Section B (9 marks) will be marked using mark bands. For good marks in your essay your response will need to contain:

- substantial evidence of sound knowledge and understanding of ESS issues and concepts
- a wide breadth of statements that show clear knowledge, effectively linked with each other and to the context of the question
- consistently appropriate and precise use of ESS terminology
- effective use of relevant, well-explained and original examples, where required
- examples that are subjected to well-balanced, insightful analysis
- explicit judgements or conclusions that are well supported by evidence
- conclusions that include some critical reflection.

Case studies will help you answer Paper 2 essay questions – make sure you have learned these from the course (examples are given in this book). You should use your own case studies to answer the essay questions rather than taking ideas from the resource booklet case study.

You need to plan your time for Paper 2 carefully:

- Do not spend too much time on one of the two sections.
- Plan your time *before the exam*.
- By practising past papers you will be able to work out how much time you need to take. This will vary from student to student, but here is some general advice:
  - ☐ You should be looking to spend less time on Section A (short-answer and data response) than on Section B (the two essays).
  - ☐ The essays need to be thought about carefully and planned – aim to spend a *minimum* of 35 minutes per essay but to move on if you are still on the first one after 40 minutes.

- Choose your essays carefully. Look at all sections of an essay before making your choices.
- Some students write pages on sections worth only a few marks, and then run out of time later on. Look carefully at the number of marks available for each question and adjust the amount of time you spend on that question accordingly. Writing a plan for your essays will help you.
- There are usually several sections in an essay question – make sure you answer all parts.

Essays should be subdivided into sections, not written as one long paragraph – examiners like this because it makes the paper easier to read and mark. Leave at least one line between sections of an essay for clarity, and note on your scripts if a continuation sheet has been used.

Assessment objectives

To successfully complete the course, you need to have achieved the following objectives:

1 Demonstrate knowledge and understanding of relevant:

- facts and concepts
- methodologies and techniques
- values and attitudes.

2 Apply this knowledge and understanding in the analysis of:

- explanations, concepts and theories
- data and models
- case studies in unfamiliar contexts
- arguments and value systems.

3 Evaluate, justify and synthesise, as appropriate:

- explanations, theories and models
- arguments and proposed solutions
- methods of fieldwork and investigation
- cultural viewpoints and value systems.

4 Engage with investigations of environmental and societal issues at the local and global level through:

- evaluating the political, economic and social contexts of issues
- selecting and applying the appropriate research and practical skills necessary to carry out investigations
- suggesting collaborative and innovative solutions that demonstrate awareness and respect for the cultural differences and value systems of others.

These assessment objectives are examined in the following way:

Assessment objectives	Which component addresses this assessment objective?	How is the assessment objective addressed?
1–3	Paper 1	Case study
1–3	Paper 2	Section A: short-answer questions
		Section B: two essays (from a choice of four)
1–4	Internal assessment: practical work	Individual investigation assessed using mark bands

Command terms

Command terms indicate the depth of treatment required for a given assessment statement. Objectives 1 and 2 address simpler skills; objectives 3 and 4 and relate to higher-order skills.

It is essential that you are familiar with these terms, for Papers 1 and 2, so that you are able to recognise the type and depth of response you are expected to provide.

The following tables show examples of all of the command terms, with selected questions to show how they can be used in exams. 'Advice for success' gives you hints about how to respond to the command term.

■ Objective 1

Demonstrate an understanding of information, terminology, concepts, methodologies and skills with regard to environmental issues.

Term	Definition	Sample question	Advice for success
Define	Give the precise meaning of a word, phrase, concept or physical quantity	Japanese knotweed can be described as a pioneer species. Define the term *pioneer species*. [1] (Paper 1, Nov 2012)	The glossary in the course guide is a good starting point for learning definitions
Draw	Represent by means of a labelled, accurate diagram or graph, using a pencil	For an ecosystem you have studied, draw a food chain of at least four named species. [1] (Paper 1, May 2011)	Be prepared to draw diagrams in both Papers 1 and 2; your answer will be electronically scanned, so draw the diagram or graph clearly and create labels that can be easily read. A ruler (straight edge) should be used for straight lines. Diagrams should be drawn to scale. Graphs should have points correctly plotted (if appropriate) and joined by a straight line or smooth curve
Label	Add labels to a diagram	Label a point on Figure 5 to show the likely location of the power station responsible for the thermal pollution of local waters. [1] (Paper 1, Nov 2012)	You need to be precise – in this case, the power station must be on the land!
List	Give a sequence of brief answers, with no explanation	List *three* types of solid domestic waste. [1] (Paper 1, May 2010)	A list is likely to consist of just a few words – you will not gain any credit for an explanation or a detailed description
Measure	Obtain a value for a quantity	Measure the decrease in the thickness of the ice sheet on the south coast of Greenland between 1950 and 2010. [1]	Use the scale to measure the extent of the decline
State	Give a specific name, value or other brief answer without explanation or calculation	State the term for the pattern of vegetation shown in Figure 1. [1] (Paper 1, Nov 2012)	The answer is likely to be short as well as specific

■ Objective 2

Apply and use information, terminology, concepts, methodologies and skills with regard to environmental issues.

Term	Definition	Sample question	Advice for success
Annotate	Add brief notes to a diagram or graph	The diagram below shows two biotic components of the carbon cycle: a plant and a bird. Annotate the diagram to show the inputs and outputs of carbon through photosynthesis and respiration. [2] (Paper 1, Nov 2010)	The notes should aid in the description or explanation of the diagram or graph
Apply	Use an idea, equation, principle, theory or law in relation to a given problem or issue	Apply Simpson's diversity index to work out the diversity of species in the woodland ecosystem. [2]	The formula for Simpson's diversity index will be supplied and does not need to be memorised
Calculate	Obtain a numerical answer, showing the relevant stages of working	Calculate the percentage increase in grain production between 1967 and 2005. [1] (Paper 1, Nov 2012)	You should include all the steps involved in calculating the answer The final response should be made clear and have appropriate units where applicable
Describe	Give a detailed account	Describe one other system where human activities have created environmental problems through a positive feedback system and explain how the system can be brought back to balance. [8] (Paper 2, Nov 2012)	Be guided by the number of marks assigned to the question; more marks require a more detailed description
Distinguish	Make clear the differences between two or more concepts or items	Distinguish between *negative feedback* and *positive feedback*. [2] (Paper 1, May 2010)	In this type of question it is essential that you emphasise the differences between the two concepts – it is not acceptable to just define the two terms/give two separate descriptions
Estimate	Obtain an approximate value	Estimate the area covered by national parks in Borneo. [1]	Use the scale on the map and your ruler to work out an approximate value You do not need to spend a lot of time measuring round curves
Identify	Provide an answer from a number of possibilities	Identify two limiting factors affecting the vegetation in the alpine meadows. [2] (Paper 1, Nov 2012)	Only a very brief answer is needed here
Outline	Give a brief account or summary	Outline *two* problems caused by the use of herbicides to control invasive species such as Japanese knotweed. [2] (Paper 1, Nov 2012)	Two negative impacts of the use of herbicides must be described briefly

■ Objectives 3 and 4

Synthesise, analyse and evaluate research questions, hypotheses, methods and scientific explanations with regard to environmental issues.

Using a holistic approach, make reasoned and balanced judgements using appropriate economic, historical, cultural, socio-political and scientific sources.

Articulate and justify a personal viewpoint on environmental issues with reasoned argument while appreciating alternative viewpoints, including the perceptions of different cultures.

Term	Definition	Sample question	Advice for success
Analyse	Break down in order to bring out the essential elements or structure	Analyse the population structure shown in the population pyramid. [4]	From within the population pyramid determine the basic age and gender demographics of the population and highlight them within your response
Comment	Give a judgement based on a given statement or result of a calculation	Comment on the relationship between population growth and food supply. [3]	You should reach a conclusion on whether you think population growth is out-stripping food resources
Compare and contrast	Give an account of similarities and differences between two (or more) items or situations, referring to both (or all) of them throughout	Compare and contrast the abiotic factors found in the straight part of the river with the abiotic factors found in the meandering part of the river. [3] (Paper 2, Nov 2012)	You should describe and explain the similarities (compare), as well as describe and explain the differences (contrast) between the abiotic factors in the two named parts of the river
Construct	Display information in a diagrammatic or logical form	Construct a simple diagram to show *three* inputs that can lead to increases in *two* outputs in a food production system. [3] (Paper 1, Nov 2012)	This may involve boxes and arrows, for example in the form of a flow diagram It is very important that your construction is clear
Deduce	Reach a conclusion from the information given	Deduce, giving a reason, whether the figure below could represent the transfer of energy in a terrestrial ecosystem. [1] (Paper 1, May 2012)	Your answer must state the conclusion reached
Demonstrate	Make clear by reasoning or evidence, illustrating with examples or practical application	Demonstrate knowledge of use of DDT. [5]	Refer to both the anti-malarial and agricultural uses
Derive	Manipulate a mathematical relationship to give a new equation or relationship	Using the crude birth rate and crude death rate, derive the natural increase for the selected populations. [2]	Crude birth and death rates are given in rates per thousand whereas natural increase is given in rates per hundred (per cent) so a conversion/manipulation of the data must be conducted to derive the answer Check the units that are used
Design	Produce a plan, simulation or model	Design a conservation area for the protection of a named species. [5]	Design a reserve, using appropriate criteria such as size, shape, edge effects, corridors and proximity to other reserves
Determine	Obtain the only possible answer	Determine the stage of demographic transition represented by each age/sex pyramid. [2] (Paper 1, Nov 2012)	You should determine a definite stage, with no alternatives
Discuss	Offer a considered and balanced review that includes a range of arguments, factors or hypotheses; opinions or conclusions should be presented clearly and supported by appropriate evidence	With reference to all of the data, discuss the relationship between natural income and the sustainability of human activities in the Danube River delta. [4] (Paper 2, Nov 2012)	The response should contain evidenced commentary on the relationship between natural income and the sustainability of human activities in the Danube River delta You must make sure that you consider both/ all sides of the issues

Term	Definition	Sample question	Advice for success
Evaluate	Make an appraisal by weighing up the strengths and limitations	Evaluate the policies or legislation or actions that exist locally, nationally and internationally that address this issue. [7] (Paper 2, Nov 2012)	You must name the policies, legislation or actions being discussed; all three levels need to be evaluated The response should include mention of both the advantages and disadvantages of the policies, and arrive at a conclusion
Examine	Consider an argument or concept in a way that uncovers the assumptions and interrelationships of the issue	Examine different points of view regarding harvesting of a controversial species. [4]	The historical Inuit tradition of whaling versus modern international conventions is a good example
Explain	Give a detailed account, including reasons or causes	Explain why rates of net primary productivity are higher in some parts of the planet than others. [2] (Paper 1, Nov 2012)	Assume that you will need to briefly describe before you explain, unless that was required in the preceding part of the question
Justify	Give valid reasons or evidence to support an answer or conclusion	Justify your personal viewpoint on the value of international cooperation in the conservation of tropical rainforests. [2] (Paper 2, May 2010)	In determining the approximate number of pieces of evidence required in your response look to the marks assigned to the question Be specific – here you need to consider the role of international cooperation, not just conservation of rainforests
Predict	Give an expected result	Predict the effect on nutrient cycling of increased precipitation over many years in a region that is currently a steppe. [3] (Paper 1, Nov 2011)	There are 3 marks available so you should refer to storages in the biomass, litter and soil, as well as the flows between them and the inputs and outputs to and from the system When you are asked to make a prediction you will be provided with information to assist you – be sure to use it
Sketch	Represent by means of a diagram or graph (labelled as appropriate)	Sketch two diagrams to show the population pyramids for LEDCs and MEDCs. [4]	The sketches should give a general idea of the required shape or relationship, and should include relevant features
Suggest	Propose a solution, hypothesis or other possible answer	Suggest *one* way in which the pattern of vegetation shown in Figure 1 might change as a result of global warming. [1] (Paper 1, Nov 2012)	The term 'suggest' is used when there are several possible answers and you may have to give reasons or a judgement
To what extent	Consider the merits or otherwise of an argument or concept	To what extent is an isolated system a useful hypothetical concept? [3]	Opinions and conclusions should be presented clearly and supported with appropriate evidence and sound argument

Countdown to the exams

4–8 weeks to go

- Start by looking at the syllabus and make sure you know exactly what you need to revise.
- Look carefully at the checklist in this book and use it to help organise your class notes and to make sure you have covered everything.
- Work out a realistic revision plan that breaks down the material you need to revise into manageable pieces. Each session should be around 25–40 minutes with breaks in between. The plan should include time for relaxation.
- Read through the relevant sections of this book and refer to the expert tips, common mistakes, key definitions, case studies and worked examples.
- Tick off the topics that you feel confident about, and highlight the ones that need further work.
- Look at past papers. They are one of the best ways to check knowledge and practise exam skills. They will also help you identify areas that need further work.
- Try different revision methods, for example summary notes, mind maps and flash cards.
- Test your understanding of each topic by working through the **Quick check** questions and **Exam practice questions** at the end of each section.
- Make notes of any problem areas as you revise, and ask a teacher to go over them in class.

1 week to go

- Aim to fit in at least one more timed practice of entire past papers, comparing your work closely with the mark scheme.
- Examine the checklist carefully to make sure you haven't missed any of the topics.
- Tackle any final problems by getting help from your teacher or talking them over with a friend.

The day before the examination

- Look through this book one final time. Look carefully through the information about Paper 1 and Paper 2 to remind yourself what to expect in both papers.
- Check the time and place of the exams.
- Make sure you have all the equipment you need (e.g. extra pens, a watch, tissues). Make sure you have a calculator – this is needed for both papers.
- Allow some time to relax and have an early night so you are refreshed and ready for the exams.

My exams

ESS Paper 1

Date:..................................

Time:..................................

Location:..............................

ESS Paper 2

Date:..................................

Time:..................................

Location:..............................

1 Foundations of environmental systems and societies

1.1 Environmental value systems

> **SIGNIFICANT IDEAS**
> - Historical events, among other influences, affect the development of environmental value systems (EVSs) and environmental movements.
> - There is a wide spectrum of EVSs, each with its own premises and implications.

Development of the modern environmental movement

Key historical events have influenced the development of the modern environmental movement. Significant historical influences on the development of the environmental movement have come from literature, the media, major environmental disasters, international agreements and technological developments. Major landmarks include:

- Minamata:
  - □ In 1956 a chemical company released toxic methyl mercury into waste water in Minamata, Japan.
  - □ Shellfish and fish were contaminated.
  - □ Local people developed illnesses caused by mercury poisoning.
  - □ This raised awareness of threats posed by industrialisation.
- Rachel Carson's *Silent Spring* (1962) raised awareness of the threat of the pesticide DDT to organisms higher up food chains (Figure 1.1).
- The Save the Whale campaign (1970s) was coordinated by Greenpeace, which organised direct action to prevent whaling.
- Bhopal:
  - □ In 1984 an explosion at a Union Carbide plant in Bhopal, India released 42 tonnes of toxic methyl isocyanate gas.
  - □ Between 8000 and 10 000 people died within the first 72 hours.
  - □ Highlighted to the world how dangerous factories can be.
- The Chernobyl nuclear meltdown in 1986 reinforced negative perceptions of nuclear power in society.
- The UN **Rio Earth Summit** (see also page 82):
  - □ The first UN conference to focus on sustainable development.
  - □ Attended by 172 nations, so its impact was likely to be felt globally.
  - □ The summit's message was radical, i.e. that nothing less than a transformation of our attitudes and behaviour would bring about the necessary changes.
  - □ Led to the adoption of **Agenda 21**, a blueprint for action to achieve sustainable development worldwide.
- *An Inconvenient Truth*:
  - □ Extensive publicity following release of the film in 2006 meant that many people heard about global warming.
  - □ A mainstream political figure championed environmental issues for the first time.
  - □ The film made the arguments about global warming accessible to a wider audience.
  - □ The film changed people's attitudes and raised awareness about climate change.

Figure 1.1 Rachel Carson – author of *Silent Spring* and pioneer of the environmental movement

> **Expert tip**
>
> You need to be able to justify, using examples and evidence, how historical influences have shaped the development of the modern environmental movement.

> **Expert tip**
>
> You do not need to remember all of the historical events that have shaped the environmental movement: questions usually ask you to discuss two or three different events. You will need to explain in detail why each one is important.

Each of these:

■ resulted in the creation of environmental pressure groups, both local and global
■ promoted the concept of stewardship
■ increased media coverage, which raised public awareness (Figure 1.2).

Attitudes towards the environment change over time:

■ When a new resource or product is first developed, people are more likely to see benefits than potential problems, which emerge later (e.g. the car).
■ Key events prompt change (see above).
■ Environmental pressure groups help to raise awareness by distributing leaflets and staging events (e.g. Greenpeace).
■ Environmental attitudes can become politically mainstream when economic consequences of pollution are seen (e.g. the Stern report on global warming).
■ School curricula can reflect and promote changing attitudes.
■ Changing technologies can help to spread new attitudes (e.g. the internet).
■ International organisations (e.g. the United Nations Environment Programme – UNEP) can raise the profile of environmental issues through conferences. These can set targets that take effect through national government strategies (e.g. Agenda 21 – page 82).

Figure 1.2 Literature on ecological issues has influenced the environmental movement

Common mistake

There is a tendency in exams to write over-long answers. If you are asked, for example, to describe the role of historical influences in the environmental movement, make sure that you give a detailed account appropriate to the number of marks awarded. If 2 marks are awarded then two different points, no more, are needed.

> ### ■ QUICK CHECK QUESTIONS
>
> 1 Describe the role of any *four* named historical influences in the environmental movement.
> 2 Outline how major landmarks have influenced public perception of environmental issues.
> 3 Describe how attitudes towards the environment can change over time.

Social systems

`Revised`

Environmental value systems (Figure 1.3), like all systems, have inputs and outputs:

■ Outputs of an environmental system are determined by processing the inputs.
■ Outputs can be modified by personal characteristics (e.g. age, gender, strong-willed vs compliant, optimistic vs pessimistic) and emotions.

> ### Keyword definitions
>
> **Ecosystem** – A community and the physical environment with which it interacts.
>
> **Social system** – People, groups and institutions that work together, forming distinct patterns and relationships that define the society.

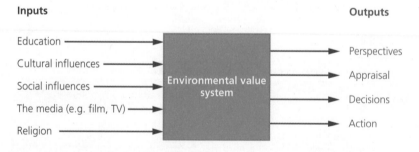

Figure 1.3 Environmental value systems: inputs and outputs

Environmental value systems work within **social systems**. Both social systems and **ecosystems** exist at different scales, and have common features such as feedback and equilibrium.

Table 1.1 Comparing social systems and ecosystems

	Social system	Ecosystem
Flows	Information, ideas and people	Energy and matter
Storage	Ideas and beliefs	Biomass; the atmosphere; soils; lakes, rivers, sea
Levels	Social levels (i.e. hierarchy)	Trophic levels
Producers	People responsible for new input (e.g. ideas, books, films)	Plants, algae and some bacteria
Consumers	Absorb and process new input (e.g. ideas, food, material possessions)	Consume other organisms

Range of environmental value systems

Revised

An environmental value system (EVS) determines the global perspective of an individual or group of individuals, the decisions they make and the course of action they take regarding environmental issues. There is a range of different EVSs:

Ecocentrism is a nature-centred EVS. An ecocentrist worldview sees nature as having an inherent value:

- Involving minimum disturbance of natural processes.
- Combining spiritual, social and environmental aspects.
- Aiming for sustainability for the whole Earth.
- Involving self-imposed restraint of natural resource use.

Anthropocentrism is a people-centred environmental EVS. It believes that it is important for everyone in **society** to participate in environmental decision making:

- People act as the managers of sustainable global systems.
- People can sustainably manage the global system through the use of taxes, environmental regulation and legislation.
- Management requires strong regulation by independent authorities.
- Participation has an important role to help educate people about environmental issues.
- Participation means that people who might be causing the problems are less likely to do so if they are involved in decisions about their own environment.
- Sometimes less powerful groups in society have the best knowledge about what is right for an environment, for example indigenous groups.
- People who believe in democracy argue that everyone has a right to have a say in how communal natural resources are managed.
- Ecosystems need to be managed holistically so people from all walks of life should be able to contribute.
- Debate is encouraged to reach a consensual, pragmatic approach to solving environmental problems.
- It includes **self-reliant, soft ecologists** who:
  - ☐ are more conservative about environmental problem solving than deep ecologists (see page 4)
  - ☐ emphasise smallness of scale and therefore community identity in work and leisure
  - ☐ lack faith in modern, large-scale technology
  - ☐ believe that materialism for its own sake is wrong
  - ☐ believe that economic growth can be geared to even the poorest people.
- It includes **environmental managers** who:
  - ☐ believe that economic growth and resource exploitation can continue if carefully managed
  - ☐ believe that laws and regulation can manage natural resources
  - ☐ appreciate that preserving biodiversity can have economic and ecological advantages
  - ☐ believe in compensation for those who experience adverse environmental or social effects.

Technocentrism is a technology-centred environmental EVS. A technocentrist worldview sees technology as providing solutions to environmental problems even when human effects are pushing natural systems beyond their normal boundaries:

- Technology can keep pace with, and provide solutions to, environmental problems.
- Resource replacement can reduce resource depletion.
- There is a need to understand natural processes in order to control them.
- Emphasis should be on scientific research and prediction before policymaking.
- Emphasis should be on sustained market and economic growth.

Specific groups representing different EVSs lie on the spectrum from ecocentrism through to technocentrism, with **deep ecologists** and **cornucopians** at opposite ends of the environmental values system continuum (Table 1.2).

Table 1.2 Comparing the environmental value systems of deep ecologists and cornucopians

	Deep ecologist	Cornucopian
Role of nature	See humans as subject to nature, not in control of it	Nature is there to be made use of by humanity
	It is of intrinsic importance for the existence of humanity	Humans can control their environment
Role of humanity	More value placed on nature than on humanity	Humans have the ability to improve the conditions of the Earth's peoples and they have the ingenuity to overcome any difficulties
	Ecological laws dictate human morality	
Biodiversity	Believe in the inherent right to life and intrinsic value of species	See biodiversity as a resource to be exploited for economic gain
	Place most value on bio-rights, i.e. favour the rights of species to remain free from harm over the rights of humans who wish to exploit resources for economic gain	Less concern for intrinsic or ethical rights of biodiversity
Consumption	Nature is more important than material gain for its own sake	Resources are there to be exploited and to generate income
Modern, large-scale technology	Distrust and lack faith in it	See it as providing the solutions to humanity's environmental issues
Economic growth	Should not occur at the expense of natural resources and the environment	Forms the basis of all projects and policies
	Should be geared to providing for the needs of the poorest people	
Environmental problems	Need to be tackled at source and ideally prevented in the first place	People can always find solutions to political, scientific or technological difficulties

Expert tip

You may be asked to comment on how people's different EVSs (e.g. deep ecologists vs cornucopians) influence how they respond to environmental issues covered in the course (e.g. energy supply, water shortage, farming methods), and how this affects decision-making processes.

Expert tip

There are extremes at either end of the EVS spectrum, from deep ecologists (ecocentric) to cornucopian (technocentric) but, in practice, EVSs vary greatly depending on cultures and time periods, and they rarely fit simply or perfectly into any classification.

Common mistake

Take care not to oversimplify the views of ecocentrism and technocentrism by making claims such as 'ecocentrics don't believe in technology' and 'technocentrics don't care about the environment'. You need to use the level of detail shown in Tables 1.2 and 1.3.

Decision making on environmental issues

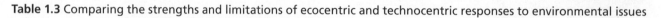

Revised

Table 1.3 Comparing the strengths and limitations of ecocentric and technocentric responses to environmental issues

	Strengths	Limitations
Ecocentrism	Approaches are more sustainable	Conservation can be costly, with little economic return
	Responses aim to minimise impact on the environment by encouraging restraint	May be unpopular with countries seeking economic development
	Promotes methods more in harmony with natural systems	May hinder economic growth and development
	Does not have to wait for technological developments to occur	Requires individual change, which can be difficult to encourage
	Raises general environmental awareness in communities	
Technocentrism	Provides alternatives that don't inconvenience people	Technological solutions may give rise to further environmental problems
	Substitutes materials and so avoids costly industrial change	Substituting does not solve the problem of consumerism
	Allows economic, social and technological development to continue	Allows for greater resource consumption
		High costs

Worked example

Describe and evaluate ecocentric and technocentric responses to eutrophication.

ECOCENTRIC

- Encourage methods that are in balance with natural systems – for example, use animal dung on agricultural fields or crop rotation, so that less or no fertiliser is needed.
- Encourage people to use less detergent through educational campaigns to promote more environmentally friendly detergent (i.e. zero or low phosphate).
- Plant a buffer zone between fields and water courses to absorb any runoff from fields treated with fertiliser.

TECHNOCENTRIC

- Promote the use of alternative materials such as alternatives to phosphates in detergents.
- Apply fertilisers more carefully so there is reduced runoff.
- Use technology to screen water and to remove pollutants – for example, phosphate stripping.
- Pump air through lakes to avoid the low-oxygen conditions.
- Remove nutrient-enriched sediments from water courses – for example, through mud pumping.

EVALUATION

- Ecocentric approaches are difficult to enforce as people are reluctant to change their lifestyles.
- Alternative approaches such as organic fertilisers may not work effectively and can still result in eutrophication.
- Technocentric solutions might increase costs – for example, research to develop new detergents requires financial commitments. They do not get to the main cause of the problem.
- Technocentric solutions provide a short-term solution but are unsustainable in the long term. They may not be an option in LEDCs.
- Technology can help to provide solutions only if there is a political will and commitment from people to make changes.

Common mistake

A question may ask you to discuss two contrasting environmental problems. The contrast can be one of cause or scale. If you do not pick two very different environmental issues you will lose marks. Appropriate contrasting issues would be, for example, climate change and biodiversity loss.

Expert tip

If asked to evaluate a response to an environmental problem, an appropriate answer might be: technology is a tool that cannot on its own solve any problem, there has to be political will (anthropocentric involvement and commitment from people) to make changes and then technology can help to provide solutions; people are reluctant to adapt lifestyles or accept lowered standards of living, so ecocentric approaches can be hard to enforce.

■ QUICK CHECK QUESTIONS

4   Define the term *environmental value system*.
5   Compare and contrast social systems and ecosystems.
6   Outline the range of environmental value systems.
7   Describe the views of an environmental manager.
8   Compare and contrast the environmental views of a deep ecologist and a cornucopian.

Environmental value systems of different societies

Revised ☐

Expert tip

You need to be able to evaluate the implications of two contrasting EVSs in the context of given environmental issues. The societies chosen should demonstrate significant differences.

When comparing environmental value systems, suitable contrasting societies would be, for example: Native Americans and European pioneers operating frontier economics, which involved exploitation of seemingly unlimited resources; Buddhist and Judaeo-Christian societies; communist and capitalist societies; shifting cultivators and urban societies.

Worked example

Compare the attitudes of *two named* contrasting societies towards the natural environment, and discuss the consequences of these attitudes to the way in which natural resources are used.

Example used: indigenous farmers using shifting cultivation in the Amazonian rainforest in Brazil, and city dwellers in Brasilia.

INDIGENOUS FARMERS

- Natural resources are used in a way that minimises impact on the environment.
- Attitudes can broadly be termed 'ecocentric' (see details in Table 1.3).
- Lifestyles and practices are compatible with the forest in which they live – using forest materials to make their homes and canoes, and for medicines.
- Use farming methods that mimic forest structure, for example by maintaining the layered structure of rainforest to protect ground crops from the Sun and heavy downpours.
- Only return to farmed sites after around 50 years to allow soil fertility to be restored.
- They are animists and recognise the spiritual role of the forest, which leads to respect for trees and other species.
- Overall they are less destructive and have a closer connection between social systems and ecological systems.

CITY DWELLERS FROM BRASILIA

- They see rainforest as a resource to be exploited for economic gain, and underestimate the true value of pristine rainforest.
- Attitudes can broadly be termed 'technocentric' (see details in Table 1.3).
- They have a lack of understanding about how natural systems work, which means that they may support decisions that lead to damaging actions (e.g. construction of dams, which then become silted up).
- They may migrate to make use of deforested land, but are unsuccessful as the soil lacks fertility.

Personal viewpoints on environmental issues

Revised ☐

Figure 1.4 outlines factors that determine personal environmental value systems. These principles can be applied to any of the environmental issues covered by the course. The EVS of an individual will be shaped by the cultural, economic and socio-political context.

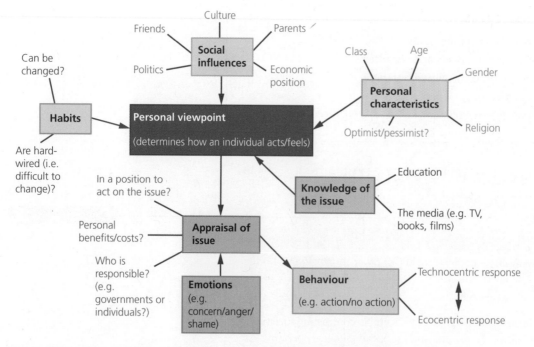

Figure 1.4 Factors that determine personal environmental value systems

■ QUICK CHECK QUESTIONS

9 Compare and contrast the environmental value systems of *two* contrasting societies.

10 What factors can determine your personal viewpoint on environmental issues?

Expert tip

The course requires you to reflect on where you stand on the spectrum of environmental value systems, and to develop your own EVS. You need to be able to justify your decisions on environmental issues based on your EVS, and consider how your viewpoint relates to all the environmental issues that you study throughout the course, such as population control, resource exploitation, sustainable development and climate change.

Expert tip

EVSs are individual; there is no 'wrong' EVS.

Intrinsic value

Revised ☐

Natural systems have different values. Humans gain value from goods and services provided by ecosystems (Chapter 8, page 154–5), most often by people visiting or residing in the ecosystem. Value also can come from ecosystems from harvesting food products, timber for fuel or housing, medicinal products and hunting animals for food and clothing. Natural systems also have **intrinsic value**.

Different EVSs ascribe different intrinsic values to components of the **biosphere** (the living part of the Earth).

Ecological values have no formal market price: soil erosion control, nitrogen fixation and photosynthesis are all essential for human existence but have no direct monetary value, although some estimates have been made.

Ecosystems that are valued on intrinsic grounds may not provide identifiable goods or services, and so remain un-priced or undervalued from an economic viewpoint.

There are many examples of places or ecosystems that have an important national identity – for example, Mount Fuji in Japan or Mount Kilimanjaro in Tanzania. Uluru (Ayers Rock) in Australia has great spiritual value for the Aboriginal population. Such areas or ecosystems have intrinsic value from an ethical, spiritual or philosophical perspective, and are valued regardless of their potential use to humans.

Intrinsic values may vary between different EVSs (see the worked example on page 6). They include values based on culture, aesthetics and bequest significance (i.e. their value to children and grandchildren).

Keyword definition

Intrinsic value – A characteristic of a natural system that has an inherent worth, irrespective of economic considerations, such as the belief that all life on Earth has a right to exist.

Expert tip

How can we quantify values such as intrinsic value, which are inherently qualitative? Have an example ready – such as Mount Fuji or Uluru – to support any statements that you make about the intrinsic value of nature in an examination.

EXAM PRACTICE

1 Identify *three* landmarks in the development of the modern environmental movement, and justify why each one was significant. [9]

2 Compare the characteristics of ecosystems and social systems. [5]

3 Compare the likely views of a deep ecologist and a cornucopian on the exploitation of coal reserves in an undisturbed ecosystem. [5]

4 Outline the arguments that might be given for preserving biodiversity by a deep ecologist and an environmental manager. [4]

5 Justify your personal viewpoint of the environmental issue shown in the photograph (right), which shows deforestation in a tropical rainforest. [6]

1.2 Systems and models

Revised ☐

SIGNIFICANT IDEAS

- A systems approach can help in the study of complex environmental issues.
- The use of systems and models simplifies interactions but may provide a more holistic view without reducing issues to single processes.

Concept and characteristics of systems

Revised ☐

A systems approach is a way of visualising complex sets of interactions. These interactions produce the emergent properties of the system.

- A system can be divided into parts, or components, which can each be studied separately: this is called a '**reductionist**' approach.
- A system can also be studied as a whole, and patterns and processes described for the whole system: this is called a '**holistic**' approach.

The course focuses on ecosystems, but the systems approach can equally be applied to economic, social and value systems. The systems approach is shown in Figure 1.5.

> **Keyword definition**
>
> **System** – An assemblage of parts and the relationships between them, which together constitute an entity or whole.

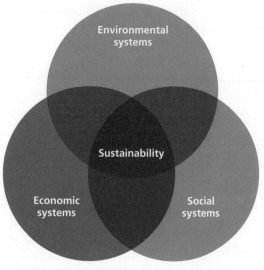

Figure 1.5 The systems approach allows different disciplines to be studied in the same way, and for links to be made between them

Figure 1.6 A simple systems diagram, showing storage and flows

Systems can be shown as diagrams consisting of storages and flows (Figure 1.6).

- **Storages** are represented by boxes.
- **Flows** are represented by arrows.
- Arrows represent **inputs** and **outputs** from the system.

An example is shown in Figure 1.7.

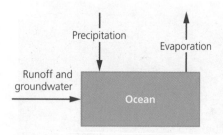

Figure 1.7 A diagram of an ocean system, with flows into and out from the storage (the ocean)

> **Expert tip**
>
> You need to be able to construct a system diagram from a given set of information.

> **Expert tip**
>
> A systems approach should be taken for all the topics covered in the ESS course.

> **Common mistake**
>
> If you are asked to construct a *diagram* of a system, do not draw a picture. This reduces the time available for completing the question. You are expected to draw diagrams with boxes and arrows, representing storages and flows. Draw bold, clear, well-labelled diagrams.

Transfer and transformation processes

Inputs into and outputs from systems can be transfer or transformation processes.

- **Transfers** are processes that involve a change in location within the system but no change in state, for example water flowing from groundwater into a river.
- **Transformations** lead to the formation of new products (e.g. photosynthesis, which converts sunlight energy, carbon dioxide and water into glucose and oxygen) or involve a change in state (e.g. water evaporating from a leaf into the atmosphere).

Storages and flows can be drawn in proportion (i.e. to scale). This **quantitative** way of representing the system gives extra information and adds value to models (Figure 1.8).

> **Common mistake**
>
> Do not confuse **transfers** with **transformations**. Transfers only involve a change in location, whereas transformations involve a change in state or new products.

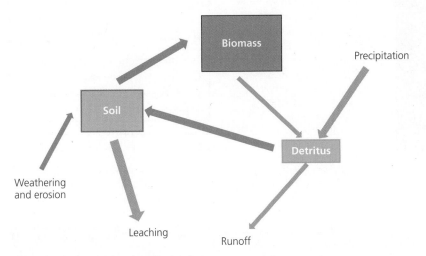

Figure 1.8 Diagram showing a nutrient model for a rainforest ecosystem. The size of the boxes represents the amount of nutrients stored, and the width of the arrows the size of the nutrient flows

> **Expert tip**
>
> If a question gives data on the size of the flows or storages, you are expected to show these on diagrams either by drawing boxes and arrows proportionally, or by including numbers.

> **Expert tip**
>
> When drawing a diagram, include processes on the input and output arrows to show the transfers (blue arrows in Figure 1.8) and transformations (red arrows in Figure 1.8) taking place.

The systems concept on a range of scales

Systems can range in size from small (e.g. a cell) to large (e.g. a rainforest). The Earth itself can be seen as a global ecosystem, with specific energy inputs from the Sun and with particular physical characteristics.

Different ecosystems exist where different species and physical or climatic environments are found. An ecosystem may, therefore, be of any size up to global. A tropical rainforest, for example, contains lots of small-scale ecosystems, such as the life found in bromeliad in the forest canopy:

- Bromeliads are flowering plants.
- They capture rainwater, which enables an ecosystem to exist that can contain tree frogs, salamanders, flatworms, snails and crabs.
- Some of these animals may spend their entire lives inside the plant.

> **■ QUICK CHECK QUESTIONS**
>
> 11 Explain the advantages of a systems approach.
>
> 12 Explain the differences between transfers and transformations. Give examples of both.

Open systems, closed systems and isolated systems

There are three different types of system – **open**, **closed** and **isolated**. Differences depend on how matter and energy move into and out of the system.

Figure 1.9 The Earth – an example of a closed system

Expert tip

When discussing whether systems are open, closed or isolated, you need to show an appropriate understanding of the roles of energy and matter, and whether they are inputs and outputs from the system.

Keyword definitions

Open system – A system that exchanges both matter and energy with its surroundings (e.g. an ecosystem).

Closed system – A system that exchanges only energy but not matter with its surroundings (e.g. the Earth – Figure 1.9).

Isolated system – A system that does not exchange either matter or energy with its surroundings (e.g. the Universe).

■ QUICK CHECK QUESTION

13 Outline the differences between open, closed and isolated systems. Give examples of each.

Models

A **model** can be used to understand how a system works and to predict how it will respond to change. All models have strengths and limitations, as shown in Table 1.4.

Expert tip

A model inevitably involves some approximation and therefore loss of accuracy.

Table 1.4 The strengths and limitations of models

Strengths	Limitations
• They simplify complex systems and allow predictions to be made.	• They might not be accurate and can be too simple.
• Inputs can be changed to see their effects and outputs, without having to wait for real events.	• They rely on the level of expertise of the people making them.
• Results can be shown to other scientists and to the public. Models are easier to understand than detailed information about the whole system.	• Different people can interpret them in different ways.
	• They may be used politically.
	• They depend on the quality of the data that go into the inputs.
	• Different models can show different outputs even if they are given the same data.

Keyword definition

Model – A simplified version of a system. It shows the flows and storages as well as the structure and workings.

Expert tip

You need to be able to evaluate the use of models as a tool in a given situation, for example, climate change predictions.

EXAM PRACTICE

6 a State what type of system the Earth is and what the inputs and
outputs are. [3]

b Evaluate the usefulness of a global perspective for managing
climate change effectively. [4]

c Scientists use computer simulations to model the effects of
climate change. Define the term model. [1]

d Discuss the advantages and disadvantages of climate change
models, and why there is uncertainty about predictions concerning
global warming. [5]

Expert tip

You need to be able to construct
a model from a given set of
information.

■ QUICK CHECK QUESTION

14 Describe the advantages and
disadvantages of models.

1.3 Energy and equilibria

Revised ☐

SIGNIFICANT IDEAS

- The laws of thermodynamics govern the flow of energy in a system and the
ability to do work.
- Systems can exist in alternative stable states or as equilibria between which
there are tipping points.
- Destabilising positive feedback mechanisms will drive systems towards these
tipping points, whereas stabilising negative feedback mechanisms will resist
such changes.

The first and second laws of thermodynamics and their relevance to environmental systems

Revised ☐

The first law of thermodynamics (law of conservation of energy) states that energy
entering a system equals energy leaving it, i.e. energy can neither be created nor
destroyed.

The second law of thermodynamics states that energy in systems is gradually
transformed into heat energy due to inefficient transfer, thereby increasing
disorder (**entropy**).

- Energy flows through ecosystems. Energy enters as sunlight energy and is
converted to new biomass and heat.
- The energy entering the system equals the energy leaving it (first law).
- Energy is inefficiently moved through food chains in the process of respiration
and production of heat energy (second law).
- Initial absorption and transfer of energy by producers is also inefficient due to
reflection, transmission, light of the wrong wavelength and inefficient transfer
of energy in photosynthesis (second law).
- Light energy starts the food chain but is then transferred from producer to
consumers as chemical energy.
- As a result of the inefficient transfer of energy, food chains tend to be short.

Keyword definition

Entropy – A measure of the
amount of disorder, chaos or
randomness in a system; the
greater the disorder, the higher the
level of entropy.

Expert tip

You need to be able to explain
the implications of the laws of
thermodynamics to ecological
systems.

The nature of equilibria

Revised ☐

In ecosystems, such as temperate forests, inputs and outputs of energy and matter
change over time. This leads to changes in the population dynamics of communities,
with populations increasing and decreasing in abundance. Overall the forest remains
the same (**steady-state equilibrium**). Systems have a tendency to return to the
original equilibrium, rather than adopting a new one, following disturbance.

Systems can have **stable** or **unstable** equilibrium (see Figures 1.10, 1.11, 1.12 and 1.13).

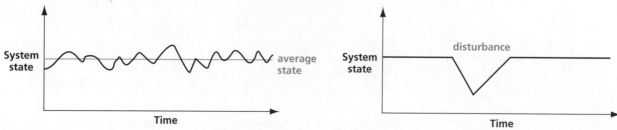

Figure 1.10 Steady-state equilibrium. There are continuing inputs and outputs of matter and energy, but the system as a whole remains in a more-or-less constant state (for example, a climax ecosystem)

Figure 1.11 Stable equilibrium. Disturbance leads to a return to the original equilibrium

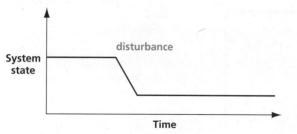

Figure 1.12 Unstable equilibrium. The system moves to a new equilibrium following disturbance

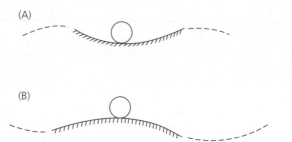

Figure 1.13 Diagrams showing the difference between (A) stable and (B) unstable equilibrium

> ### Keyword definitions
>
> **Equilibrium** – A state of balance among the components of a system.
>
> **Steady-state equilibrium** – The condition of an open system in which there are no changes over the longer term, but in which there may be oscillations in the very short term.
>
> **Stable equilibrium** – The tendency in a system for it to return to a previous equilibrium condition following disturbance. This is in contrast to **unstable equilibrium**, which forms a new equilibrium following disturbance.

> ### ■ QUICK CHECK QUESTIONS
>
> **15** Define *steady-state equilibrium*.
> **16** Explain the differences between stable and unstable equilibrium.

Positive feedback and negative feedback

Revised ▢

Figures 1.14 and 1.15 show examples of positive and negative feedback.

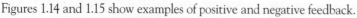

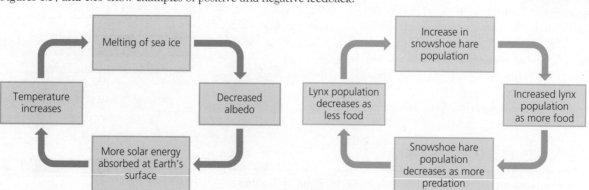

Figure 1.14 An example of positive feedback: the effect of rising temperatures on planetary albedo

Figure 1.15 An example of negative feedback: the predator–prey relationship between snowshoe hare and lynx in the boreal forest of North America

- All **feedback** links involve time lags.
- **Positive feedback** tends to amplify change away from equilibrium (Figure 1.16), whereas **negative feedback** mechanisms help to maintain stability (Figure 1.17).

Figure 1.16 Arctic ice melting, leading to positive feedback through decreased planetary albedo

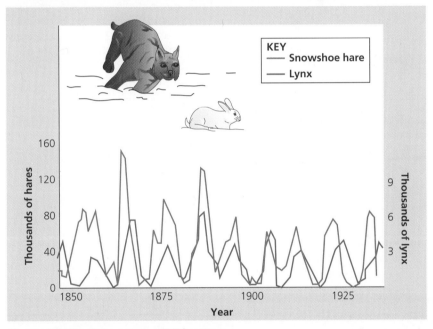

Figure 1.17 Predator–prey relationships

Expert tip

Use case studies you have studied to illustrate examples of positive and negative feedback. For example, when discussing climate change you could use: increased temperatures lead to increased melt of permafrost, increasing release of methane, increasing temperatures (positive feedback); increased carbon dioxide leads to increased plant productivity, leading to increased growth, resulting in reduced carbon dioxide (negative feedback).

Common mistake

The term 'negative' does not mean that the feedback loop is detrimental to the environment. Quite the opposite – it usually counteracts deviation away from steady-state equilibrium.

The term 'positive' does not mean that the feedback loop has a constructive effect on the environment. Positive feedback increases change in a system, leading to it moving further away from steady-state equilibrium.

Tipping points

Tipping points occur when there is a dramatic change in the ecological state, away from equilibrium. They represent points beyond which irreversible change or damage occurs. Positive feedback loops tend to amplify changes and drive the system towards a tipping point where a new equilibrium is adopted (Figures 1.12 and 1.13B). Such changes are caused by human population growth and associated factors (Figure 1.18), such as:

- resource consumption
- habitat transformation and fragmentation
- energy production and consumption
- climate change.

> **Keyword definition**
>
> **Tipping point** – A critical threshold when even a small change can have dramatic effects and cause a disproportionately large response in the overall system.

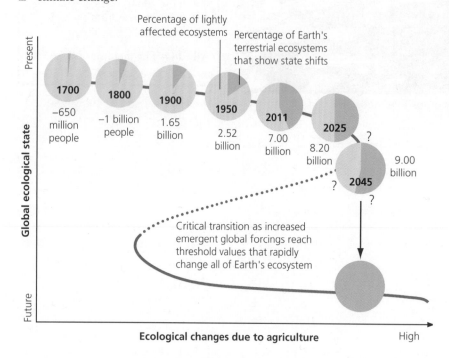

Figure 1.18 Approaching a tipping point in the Earth's biosphere. Light green indicates the fraction of Earth's land within the dynamics characteristic of the past 11 000 years; dark green indicates the fraction of terrestrial ecosystems that have undergone dramatic state changes; question marks emphasise that scientists still do not know how much land would have to be directly transformed by humans before a tipping point occurs

All of these factors exceed, in both rate and magnitude, the changes seen in the most recent global-scale shift in equilibrium at the end of the last ice age. Most projected tipping points are linked to climate change (Chapter 7).

Increases in CO_2 levels above a certain value (450 ppm) would lead to increased global mean temperature, causing melting of the ice sheets and permafrost (Chapter 7).

Tipping points caused by climate change could cause long-term damage to societies, the melting of Himalayan mountain glaciers, and a lack of fresh water in many Asian societies.

Resilience of a system

The **resilience** of an ecological or social system refers to its tendency to avoid tipping points, and maintain stability through steady-state equilibrium.

Diversity and the size of storages within systems can contribute to their resilience and affect the speed of response to change:

- Large storages and high diversity mean that a system is less likely to reach a tipping point and move to a new equilibrium.
- Complex ecosystems such as rainforests have complex food webs, which allow animals and plants many ways to respond to disturbance of the ecosystem and thus maintain stability. They also contain long-lived species and dormant seeds and seedlings that promote steady-state equilibrium.
- Agricultural crops are monocultures (i.e. containing one species) and so have low diversity and resilience, compared with naturally occurring grasslands.

> **Keyword definition**
>
> **Resilience** – The tendency of a system to avoid tipping points and maintain stability through steady-state equilibrium

> **Expert tip**
>
> The resilience of a system, ecological or social, refers to its tendency to avoid such tipping points and maintain stability.

Humans can affect the resilience of systems through reducing these storages and diversity.

■ Tropical rainforest has high diversity but disturbance through removal of tree biomass storages can lower its resilience leading to longer recovery time.

■ Nutrients are locked up in decomposing plant matter on the surface of the soil in rainforest, and in rapidly growing plants. When the forest is disturbed, nutrients are quickly lost when the leaf layer and topsoil are washed away, and when tree biomass is harvested.

■ Fires can affect grasslands and crops. Natural grasslands have high resilience, because a lot of nutrients are stored below ground in root systems, so after fire they can recover quickly. Agricultural crops are destroyed by fires because of their low resilience.

EXAM PRACTICE

7 Sabah is a Malaysian state, located in northern Borneo. The natural vegetation is tropical rainforest, and coral reef is found around the coast. Borneo has oil reserves in offshore waters.

   a Tourism is an important economic activity in Sabah. Construct a model (diagram) that demonstrates the range of impacts tourism may have both directly and indirectly on Sabah ecosystems. [4]

   b Define the term *negative feedback*. [1]

   c Define the term *positive feedback*. [1]

   d Explain why most ecosystems, such as rainforest, are negative feedback systems. [1]

   e With the help of a diagram, describe the circulation of carbon in Borneo. On your diagram describe at least *four* storages and *three* processes. [7]

8 a State, giving the reason, why natural systems are never 'isolated'. [1]

   b Draw a flow diagram model to illustrate an example of negative feedback within an ecosystem. [2]

■ QUICK CHECK QUESTIONS

17 Compare negative and positive feedback mechanisms.

18 Give *two* examples of negative feedback, and *two* of positive feedback, which are relevant to the ESS course.

1.4 Sustainability

Revised

SIGNIFICANT IDEAS

• All systems can be viewed through the lens of sustainability.
• Sustainable development meets the needs of the present without compromising the ability of future generations to meet their own needs.
• Environmental indicators and ecological footprints can be used to assess sustainability.
• Environmental impact assessments (EIAs) play an important role in sustainable development.

Resources and natural income

Revised

■ Resources are everything that is useful to mankind.

■ They include air, water, soil, people, education, fossil fuels, ecosystems and so on.

■ People can get many benefits from resources.

■ Some environmentalists describe resources as 'natural capital'.

■ They often make a comparison with saving money (capital) in a bank. At the end of the year, the savings (capital) may have gained some interest.

■ Likewise, with natural resources (**natural capital**), over time these may produce more resources, and so people can live off the 'interest' – this is known as '**natural income**'.

■ These goods and services include marketable goods such as timber and food or may be in the form of ecological services such as the flood and erosion protection provided by forests (services) and climate regulation.

■ Some of these services are impossible to quantify.

■ Non-renewable resources, such as oil and coal, generate wealth but can be used only once in a human lifetime.

Keyword definitions

Natural capital – Natural resources that are managed to provide a sustainable natural income from goods or services.

Natural income – The portion of natural capital (resources) that is produced as 'interest', i.e. the sustainable income produced by natural capital.

Ecosystem services

There are four main types of ecosystem service:

- **Supporting services** are the essentials for life and include primary productivity, soil formation and the cycling of nutrients.
- **Regulating services** are a diverse set of services and include pollination, regulation of pests and diseases and production of goods such as food, fibres and wood. Other services include climate and hazard regulation and water quality regulation.
- **Provisioning services** are the services people obtain from ecosystems such as food, fibre, fuel (peat, wood and non-woody biomass) and water from **aquifers,** rivers and lakes. Goods can be from heavily managed ecosystems (intensive farms and fish farms) or from semi-natural ones (such as by hunting and fishing).
- **Cultural services** are derived from places where people's interaction with nature provides cultural goods and benefits. Open spaces – such as gardens, parks, rivers, forests, lakes, the sea-shore and wilderness areas – provide opportunity for outdoor recreation, learning, spiritual well-being and improvements to human health.

> **■ QUICK CHECK QUESTION**
>
> **19** Identify *two* forms of natural income that might be derived from the damming of rivers.

Table 1.5 Ecosystem services

Mountains, moorlands and heaths	Woodlands
Food*	Timber*
Fibre*	Species diversity*
Fuel*	Fuelwood*
Fresh water*	Fresh water*
Climate regulation†	Climate regulation†
Flood regulation†	Flood regulation†
Wildfire regulation†	Erosion control†
Water quality regulation†	Disease and pest control†
Erosion control†	Wildfire regulation†
	Air and water quality regulation†
	Soil quality regulation†
	Noise regulation†
Recreation and tourism*	Recreation and tourism*
Aesthetic values*	Aesthetic values*
Cultural heritage*	Cultural heritage*
Spiritual values*	Employment*
Education*	Education*
Sense of place*	Sense of place*
Health benefits*	Health benefits*

Key:

Items marked * denote goods

Items marked † denote services

Items in red are considered to be from provisioning services

Items in blue are from regulating services

Items in green denote cultural services

> **Expert tip**
>
> You need to be able to discuss the value of ecosystem services to a society.

Sustainability, natural capital and natural income

■ The term **sustainability** has a precise definition (see Keyword definition).
■ Any society that supports itself in part by depleting natural capital is unsustainable. If human wellbeing is dependent on the goods and services provided by certain forms of natural capital, then long-term use rates should not exceed rates of natural capital renewal.
■ Sustainability means living within the means of nature, on the 'interest' or sustainable income generated by natural capital – for example, harvesting renewable resources at a rate that will be replaced by natural growth demonstrates sustainability.

> **Keyword definition**
>
> **Sustainability** – The use of global resources at a rate that allows natural regeneration and minimises damage to the environment.

The harvesting of timber illustrates the concept of sustainability:

■ If the rate of forest removal is less than the annual growth of the forest (i.e. the natural income), then the forest removal is sustainable.
■ If the rate of forest removal is greater than the annual growth of the forest, then the forest removal is unsustainable.
■ Sustainability focuses on the rate of resource use and suggests maintaining a balance between resource use and natural income.

> **Expert tip**
>
> You need to be able to explain the relationship between natural capital, natural income and sustainability.

Not surprisingly, much of the sustainability debate centres on the problem of how to weigh conflicting values in our treatment of natural capital.

> **■ QUICK CHECK QUESTIONS**
>
> **20** Define *sustainability*.
> **21** Explain what is meant by *natural income*. In what way is natural income sustainable?·

> **Expert tip**
>
> Deforestation can be used to illustrate the concept of sustainability and unsustainability.

Sustainable development

The term **sustainable development** was first used and defined in 1987 in *Our Common Future* (The Brundtland Report). There are many variations on the theme of sustainable development – for example, sustainable urban development, sustainable agricultural development and sustainable economic development.

To an economist, sustainable economic development might suggest economic growth remaining consistently high whereas for an environmentalist sustainable economic development might suggest using renewable energy resources to produce environmentally friendly goods.

Some critics may argue that the evolution of sustainable development has done little to change the way in which the world works. Others may argue that we are now increasingly aware of the environmental and social consequences of human impacts at a local scale and a global scale. Being aware of the inequalities allows us to plan for a better future.

■ In 1972 the **Stockholm Declaration** (the UN Conference on the Human Environment) was the first international meeting about the global environment and development.
■ In 1987 the Brundtland Commission defined the term sustainable development.
■ In 1992, at the **Rio de Janeiro Earth Summit**, **Local Agenda 21** statements were introduced. These were for all levels of government – from national to local – to help improve the environment.
■ In 1997 the **Kyoto Protocol** introduced attempts to reduce emissions of CO_2.
■ The 2002 **Johannesburg World Summit on Sustainable Development** focused on social issues, such as poverty, sanitation and access to water.
■ In 2012 the **Rio +20** conference claimed that there was a 'techno fix for every problem'. In addition, it stated that any agreements would not be legally binding.

Table 1.6 Aspects of sustainable development

Economy	Society	Environment
• Economics of sufficiency not greed • Energy-efficient buildings • Green commuting • Reduced pollution • Reduce, reuse, recycle policies	• Cultural diversity and social stability • Lifestyle and recreational amenities • Protected common land • Education and awareness • Political action for sustainability • Sustainable built environment	• Renewable energy sources • Waste management and water treatment • Reduce, reuse, recycle policies • Protected areas and wildlife corridors

> **Keyword definition**
>
> **Sustainable development –** Development that meets current needs without compromising the ability of future generations to meet their own needs.

Common mistake

Students frequently mistake *sustainability* and *sustainable development*. Remember, sustainability is the use of global resources at a rate that allows natural regeneration and minimises damage to the environment, whereas sustainable development is development that meets current needs without compromising the ability of future generations to meet their own needs.

Common mistake

Many projects that societies undertake appear sustainable. Hydroelectric power (HEP) is a good example. However, a huge amount of oil and gas is used in the building of dams and transporting materials to the dam site. Moreover, the country using the dam may well use other non-renewable forms of energy in its 'energy mix'.

Expert tip

Learn some specific examples of sustainable schemes – for example, Curitiba (Brazil) has a sustainable transport programme and a 'green exchange' whereby residents can exchange waste products for food or bus tickets.

■ QUICK CHECK QUESTIONS

22 Define *sustainable development*.
23 What were the main conclusions of the Rio +20 conference?

EXAM PRACTICE

9 Explain the relationship between natural income and sustainability. [3]
10 Distinguish between the terms *sustainability* and *sustainable development*. [4]
11 The diagram below shows features of sustainable development.

a State the *three* main aspects of sustainable development. [3]
b Outline ways in which some of these features may be linked. [3]

Millennium Ecosystem Assessment (MA)

The Millennium Ecosystem Assessment (MA) gave a scientific appraisal of the condition and trends in the world's ecosystems and the services they provide using environmental indicators, as well as the scientific basis for action to conserve and use them sustainably.

- Factors such as biodiversity, pollution, population or climate may be used quantitatively as environmental indicators of sustainability.
- Factors used to assess sustainability can be applied on a range of scales, from local to global.

The aim of the MA was to assess the consequences of ecosystem change for human wellbeing. The MA also looked at how the conservation and sustainable use of those systems could be implemented, and their contribution to human wellbeing improved. The reports produced by the MA provided a review of the conditions of the world's ecosystems and the services they provide, and the options to restore, conserve or enhance the sustainable use of ecosystems.

The main findings of the MA were as follows:

- Ecosystems have been changed more rapidly in the past 50 years than in any previous period in history, through human activities, resulting in a substantial loss in the diversity of life on Earth.
- Changes have increased the poverty of some societies.
- The problems caused by ecosystem degradation will, unless addressed, substantially reduce the benefits that future generations obtain from ecosystems.
- It will be possible to restore ecosystems but this will involve significant changes in policies and practices.
- Human actions are diminishing Earth's natural capital at a faster rate than it is being restored, which is putting pressure on natural systems and the environment.
- The ability of the planet's ecosystems to sustain future generations can no longer be taken for granted.

MA indicated that it may be possible to reverse changes as long as appropriate actions are taken quickly.

> **Expert tip**
>
> You need to be able to discuss how environmental indicators such as MA can be used to evaluate the progress of a project to increase sustainability.

Environmental impact assessments

An **Environmental impact assessments** (EIA) is carried out before a development project or large change in the way an area of land is used. It includes a **baseline study** to measure environmental conditions before development commences, and to identify areas and species of conservation importance. The baseline study is used to try to forecast what changes might be caused by the development and includes measurement of abiotic components, biodiversity, aesthetic aspects (e.g. scenery) and human populations in the area. It identifies possible impacts, predicts the scale of potential impacts and finds ways to lower the impacts.

The resulting report is known as an **environmental impact statement** (EIS) or environmental management review in some countries. A **non-technical summary** is also produced so that the general public can understand the issues.

Development projects include: road construction, hydroelectric power plants, river dams, airports, new housing, mines and brownfield development. Monitoring should continue for some time after the development. The purpose of an EIA is to weigh the advantages and disadvantages of the proposed development before the project proceeds.

- This aids in the planning of the development.
- It helps to understand the environmental impact that a project might have before it is put into place.
- It helps to determine ways to minimise the damage done to the environment.
- It helps to support the goal of sustainable development.
- It investigates economic benefits and other positive impacts of the project.

> **Keyword definition**
>
> **Environmental impact assessment –** (EIA) A method of detailed survey required, in many countries, before a major development. Ideally it should be independent of, but paid for by, the developer. It should include a baseline study, and monitoring should continue after completion of the project.

Expert tip

EIAs incorporate baseline studies before a development project is undertaken. They assess the environmental, social and economic impacts of the project, predicting and evaluating possible impacts and suggesting mitigation strategies for the project. They are usually followed by an audit and continued monitoring. Each country or region has different guidance on the use of EIAs.

Expert tip

EIAs provide decision makers with information in order to consider the environmental impact of a project. There is not necessarily a requirement to implement an EIA's proposals, and many socio-economic factors may influence the decisions made.

Expert tip

Criticisms of EIAs include: the lack of a standard practice or training for practitioners, the lack of a clear definition of system boundaries and the lack of inclusion of indirect impacts.

CASE STUDY

LONDON 2012

An EIA was carried out to assess the impacts of the London 2012 Olympics on the area of east London where the Olympic Park was to be built. The EIA was structured to address the environmental effects resulting from development of the Olympic Park. It was divided into four phases:

- Olympic and Paralympic Construction, 2007–2011
- the period during the Olympic and Paralympic Games, 2012
- Olympic Legacy Transformation, 2013–2014
- Olympic Legacy, 2015–2021.

The environmental effects were assessed for each of these phases against a baseline, which represented the conditions on the site if the London 2012 Games had not taken place.

The EIA identified the likely impacts of the development and proposed measures to reduce or offset adverse effects (mitigation measures). For example, during the construction phase, it was predicted that construction traffic would affect the flow of traffic, air quality, noise levels and the general character of the area around the site.

It was proposed that these problems would be reduced by limiting and controlling times when construction traffic was active, using a waste management strategy that limited the amount of waste being transported, and using local waterways to move material on and off the site.

A seven-point scale was used to assess the environmental impacts (Table 1.7):

1 = major adverse

2 = moderate adverse

3 = minor adverse

4 = neutral

5 = minor beneficial

6 = moderate beneficial

7 = major beneficial

Conclusions about environmental impacts assumed that proposed mitigation measures had been put into place.

Table 1.7 The overall assessments of environmental effects, using the seven-point scale

Environmental impact	Assessment period			
	2007–2011	2012	2013–2014	2015–2021
Traffic and transport:	3	2	4	4/5
highways	2/3	3/4	5/6	4/5
public transport	4/5	2/3	6	7
walking and cycling				
Energy:	3	4	6	5
energy infrastructure	4	4	4	4
energy demand	4	4	4	4
carbon emissions	4	4	5	5
heat island effect				
Socio-economic and community:	6	6/7	6	6
employment	4	7	6	7
sport and leisure	5	5	5	5
retail	6	7	7	7
culture	5	6	6	6
health				
Visual effects	2	6	2	6
Soil conditions, groundwater, contamination	7	4	3	4
Water:	3	4	3	5
water quality	3	6	6	6
aquatic ecology	2	3	3	3
hydrology	4	4	4	4
flood risk				
Terrestrial ecology and nature conservation	3	3	3	5
Air quality	4	4	4	4

Overall, by the final stage of the development, all environmental effects were predicted to be neutral or better, with major benefits for walking and cycling, sport, leisure and culture.

In May 2009 the Olympic Board stated that London 2012 had a clear policy for alleviating the impacts of manufacture, supply, use and disposal of material for the 2012 Olympic Games. The design of the Olympic Park therefore ensured that the environmental impacts were minimised.

■ **QUICK CHECK QUESTIONS**

24 Describe the stages of an environmental impact assessment.

25 Describe and evaluate the use of environmental impact assessments.

Expert tip

You need to be able to evaluate the use of EIAs.

Ecological footprints

Revised ☐

The **ecological footprint (EF)** of a population is the area of land, in the same vicinity as the population, that would be required to provide all the population's resources and assimilate its wastes. The EF is a useful model for assessing the demands that human populations make on their environment (Chapter 8).

Keyword definition

Ecological footprint (EF) – The area of land and water required to support a defined human population at a given standard of living. The measure takes account of the area required to provide all the resources needed by the population, and the assimilation of wastes.

Expert tip

You need to be able to explain the relationship between EFs and sustainability.

If the EF is greater than the area available to the population, this is an indication of unsustainability.

1.5 Humans and pollution

SIGNIFICANT IDEAS
- Pollution is a highly diverse phenomenon of human disturbance in ecosystems.
- Pollution management strategies can be applied at different levels.

The nature of pollution

Pollutants can originate from a wide range of human activities, such as the combustion of fossil fuels. They may be in the form of:

- organic or inorganic substances
- light
- sound
- thermal energy
- biological agents
- invasive species.

The impact of pollutants can be shown as systems diagrams (Figure 1.19).

> **Keyword definition**
>
> **Pollution** – The addition of a substance or an agent to an environment through human activity, at a rate greater than that at which it can be rendered harmless by the environment, and which has an appreciable effect on the organisms in the environment.

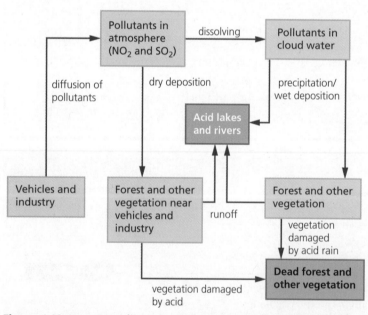

Figure 1.19 A systems diagram to show the impact of pollutants: acid rain and its effects

Expert tip

You need to be able to construct systems diagrams to show the impact of pollutants.

The major sources of pollutants

Revised ☐

Table 1.8 Main sources of pollution

Source	Type of pollution	Causes	Effects
Agriculture	Fertilisers, manure, silage	Spreading fertilisers on fields; runoff of manure and silage	Eutrophication
	Pesticides	Spraying crops	Biomagnification and bioaccumulation
	Salination	Irrigation	Accumulation of salts in soils kills plants
Manufacturing industry	Solid waste	Disposal of by-products and waste	Contaminated land, for example the Lower Lea Valley, London (Olympic Games site, 2012)
	Toxic spills and leaks	Industrial dumping and accidents	Release of toxic substances into the environment (e.g. methyl isocyanate gas, Bhopal, India)
Domestic waste	Solid domestic waste	Waste in landfill sites	Contamination of groundwater; release of methane
	Sewage	Waste from toilets; disposable nappies	Eutrophication; reduced oxygen in water; disease
Transport	Runoff from roads	Oil leakages, road drainage	Contamination of groundwater, streams and soils
Energy	Sulfur dioxide	Burning coal	Acid precipitation
	Nitrogen oxides	Formed from atmospheric nitrogen in vehicles	Acid rain, petrochemical smog (tropospheric ozone)
	Particulates (PM10s and PM2.5s)	Combustion of fossil fuels	Reduced respiratory and heart function
	Nuclear waste	Radiation leaks	Radioactive material escaping into nuclear plant and surrounding area (e.g. Fukushima-Daiichi, Japan 2011)

■ Point-source pollution and non-point-source pollution

Pollution that arises from numerous widely dispersed origins is described as **non-point-source pollution**. **Point-source pollution** arises from a single clearly identifiable site. Point-source pollution is generally more easily managed because its impact is more localised, making it easier to control emission, attribute responsibility and take legal action.

Figure 1.20 Sewage entering a river – an example of point-source pollution

Figure 1.21 A coal-powered power plant. The pollution combines with the emission from a variety of sources and is an example of non-point-source pollution

Common mistake

A pollutant only becomes a pollutant when there is too much. For example, not all fertiliser/manure causes pollution. Fertiliser that is used by plants does not cause pollution. Nor does spreading manure on fields cause pollution – as long as the amount spread is able to be utilised by plants.

Keyword definitions

Point-source pollution – The release of pollutants from a single, clearly identifiable site – for example, a factory chimney or the waste disposal pipe of a factory into a river (Figure 1.20).

Non-point-source pollution – The release of pollutants from numerous, widely dispersed origins – for example, gases from the exhaust systems of vehicles or power plants (Figure 1.21).

Expert tip

If the pollution is located at a point distant from the possible source of pollution, it is possible that many potential sources could be involved, i.e. non-point-source pollution.

■ Persistent or biodegradable pollution

Persistent pollutants cannot be broken down by living organisms and accumulate along food chains. For example, early pesticides were persistent (e.g. DDT – see page 27).

Persistent organic pollutants (POPs) are organic compounds that are resistant to environmental breakdown through biological, chemical or photolytic (i.e. broken down by light) processes.

Biodegradable means capable of being broken down by natural biological processes. Biodegradable pollutants are not stored in biological matter or passed along food chains. Most modern pesticides are biodegradable (e.g. Bt – *Bacillus thuringiensis*–proteins that are rapidly decomposed by sunlight).

■ Acute or chronic pollution

> **Keyword definitions**
>
> **Acute pollution** – Pollution that produces its effects through a short, intense exposure. Symptoms are usually experienced within hours.
>
> **Chronic pollution** – Pollution that produces its effects through low-level, long-term exposure. Disease symptoms develop up to several decades later.

Acute effects of, for example, air pollution, include asthma attacks; **chronic effects** include lung cancer, chronic obstructive pulmonary disease (COPD) and heart disease. The acute and chronic effects of exposure to UV light are examined in Chapter 6 (page 122).

■ Primary or secondary pollution

> **Keyword definitions**
>
> **Primary pollutant** – A pollutant that is active on emission.
>
> **Secondary pollutant** – A pollutant that arises from a primary pollutant that has undergone physical or chemical change.

Oxides of nitrogen (NOx) and volatile organic compounds (VOCs) are examples of **primary pollutants**, released by motor vehicle exhaust, industrial activity and chemical solvents. Ground-level ozone is an example of a **secondary pollutant**, where NOx and VOCs react with sunlight to form tropospheric (ground-level) ozone (Chapter 6, page 121).

> **■ QUICK CHECK QUESTIONS**
>
> **26** Radiation from the Fukushima-Daiichi nuclear power station is a form of which type of pollution?
>
> **27** Exhaust from cars is a form of which type of pollution?

> **Expert tip**
>
> For some pollutants there may be a time lag before an appreciable effect on organisms is evident.

Pollution management: the process of pollution and strategies for reducing impacts

Revised ☐

Pollutants are produced through human activities and create long-term effects when released into ecosystems. Strategies for reducing these impacts can be directed at three different levels in the process: altering the human activity; regulating and reducing quantities of pollutant released at the point of emission; and cleaning up the pollutant and restoring ecosystems after pollution has occurred.

Factors affecting the choice of pollution management strategy (Figure 1.22) also vary at local and national scales:

■ At a local scale, local attitudes, cultural beliefs in the environment and the enforcement of local authorities may influence the choice of pollution management strategies.
■ At a national scale, economic resources, national legislation and the political agenda may influence such choices.

Process of pollution   **Level of pollution management**

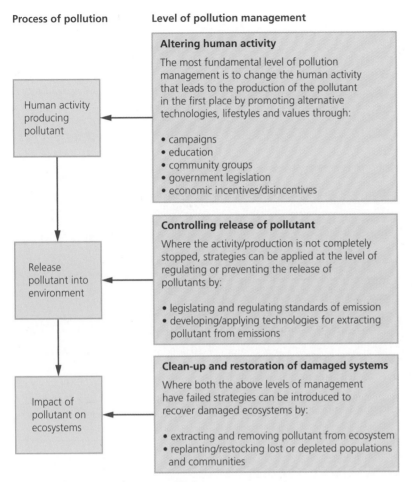

Figure 1.22 Approaches to pollution management

Expert tip

Using Figure 1.22, you should be able to show the value and limitations of each of the three different levels of intervention. In addition, you should appreciate the advantages of employing the earlier strategies over the later ones and the importance of collaboration in the effective management of pollution.

Common mistake

Some pollution can be natural – volcanic eruptions can cause acidification and climate change – so it is not entirely due to humans.

■ **QUICK CHECK QUESTIONS**

28 'Setting and imposing standards' is an example of which type of pollution management strategy?

29 How can human activities be altered to reduce the amount of pollution produced?

It is cheaper and more efficient to alter human activities. However, most action over pollution involves treating the effects of pollution (rather than altering behaviour and the causes of pollution). Treating pollution is very costly and wasteful.

Human factors that affect the approaches to pollution management

Revised ☐

▧ Cultural values

■ Capitalist societies often consider profit over the environmental damage of pollution.
■ Often they would rather treat the symptoms (effects) of pollution, when caught, rather than address the causes of pollution.
■ Rural society, with its low population density, may have an 'out of sight, out of mind' mentality, with pollution not being a problem if people don't see it or are not adversely affected by it.
■ Pollution tolerance levels vary from society to society – some countries accept the waste of other countries for recycling.
■ Some types of pollution are more tolerated than others by a particular culture – for example, noise or visual pollution in a rapidly urbanising city is accepted.
■ Cultural perspectives can be altered through education.

▧ Political systems

■ Less developed countries are often willing to allow pollution to encourage local industry – Mexico's *macquiadoras* industries are a good example.
■ The dumping of toxic waste from MEDCs to LEDCs is sometimes allowed by the governments and sometimes it is done illegally – for example in the case of the Dutch company Trafigura and the dumping of toxic waste in Côte d'Ivoire.

- Lower standards for pollution may encourage industry into certain countries. Many footwear companies in LEDCs may have dangerously high levels of glue in the workplace.
- A political blind eye may be turned if the industry is profitable, paying taxes and creating jobs.
- LEDCs often do not have the resources to enforce the laws that they do have in place.

Economic systems

- Many rich countries have a 'throwaway' society and so generate a large amount of waste and pollution.
- Increasingly in the richest countries people value a clean and tidy environment, so pollution is not tolerated.
- All three steps of the pollution model are likely to be carried out.
- In many MEDCs the most common step may be the second as the rich society might want to keep the pollution-causing industry, but regulate it.
- Poorer countries often recycle large amounts of waste through informal waste pickers – in Cairo, the Zabbaleen waste collectors recycle up to 80% of the waste they collect.
- Many LEDCs can only afford old, polluting equipment and have limited resources for technology to clean up pollution.
- In some cases UN protocols have not been signed as countries fear they may slow the economy – the USA's failure to sign up to the Kyoto Protocol is a case in point.
- As countries develop there is a tendency to spend more money on pollution prevention.

> **Expert tip**
>
> When discussing the human factors that influence pollution management strategies, real examples should be used.

> **Common mistake**
>
> Although many MEDCs invest in pollution prevention technologies, they still contribute a significant proportion of the world's pollution through travel and transport, and the import of goods produced for their benefit.

> **■ QUICK CHECK QUESTIONS**
>
> **30** Identify the population group that recycles up to 80% of the waste that it collects.
>
> **31** Suggest why an LEDC might accept waste materials from another country.

The costs and benefits to society of the WHO's ban on the use of DDT

Revised

DDT (dichlorodiphenyltrichloroethane) is a synthetic (man-made) pesticide that has many advantages and disadvantages. During the 1940s and 1950s it was used extensively to control the lice that spread typhus and the mosquitoes that spread malaria. It was also used as a pesticide in farming.

The environmental impact of DDT is based on two processes:

- **Bioaccumulation** refers to the build-up of non-biodegradable or slowly biodegradable chemicals in the body. DDT gets stored in fat tissues because it is not recognised as a toxin and is not excreted.
- **Biomagnification** refers to the process whereby the concentration of a chemical substance increases at each trophic level – the end result is that a top predator may have an accumulation that is several thousand times greater than that of a primary producer. An example of the biological effect of DDT is the thinning of eggshells in birds at the top of food chains.

In 2001 the Stockholm Convention on POPs (**persistent organic pollutants**) regulated the use of DDT – it was banned for use in farming but was permitted for disease control.

The plan is to find alternatives for disease control by 2020. In 2006 the World Health Organization (WHO) recommended the use of DDT for regular treatment in buildings in areas with a high incidence of malaria. WHO still aims for a total phase out of the use of DDT.

DDT and the battle against malaria

- The World Bank estimates that there are about 250 million cases of malaria each year.
- In South America, cases of malaria increased after countries stopped spraying DDT.

- Indoor (residual) spraying (IRS) appears to limit the growth of malarial incidence.
- However, malaria is still on the rise and many people believe that this increase in the disease is unacceptable.

DDT and the ecocentrist viewpoint

- The ecocentrist position is that ecological laws should drive human decision making.
- Chemicals such as DDT being added to the environment could harm many species or spread from houses and into the local environment.
- This affects the right of organisms to remain unmolested – for example, mosquitoes have the right to exist.
- Ecocentrists argue that alternative (less objectionable) strategies exist, such as lowering population density, removing standing water, using natural repellents and encouraging natural predators.

DDT exemplifies a conflict between the utility of a 'pollutant' and its effect on the environment.

Expert tip

You need to be able to evaluate the uses of DDT.

Expert tip

There are links between DDT and premature births, low birth weight and reduced mental development.

Common mistake

DDT does not eradicate all mosquitoes – there is evidence of growing resistance to the pesticide. Those that are resistant to DDT will continue to reproduce and spread the resistance.

■ QUICK CHECK QUESTIONS

32 What were the goals of the 2001 Stockholm Convention?

33 What do the letters IRS stand for?

EXAM PRACTICE

12 Evaluate the costs and benefits of the use of DDT to society. [6]

Topic 2 Ecosystems and ecology

2.1 Species and populations

> **SIGNIFICANT IDEAS**
> - A species interacts with its abiotic and biotic environments, and its niche is described by these interactions.
> - Populations change and respond to interactions with the environment.
> - Any system has a carrying capacity for a given species.

Species, populations, habitats and niches

Figure 2.1 A population of zebra in an African savannah

> **Expert tip**
>
> You need to be able to refer to the terms **species**, **population**, **habitat** and **niche** with reference to specific named examples. For example, valid named species should be used, such as 'Père David's deer' rather than 'deer', 'marram grass' rather than 'grass' and 'English elm' rather than 'tree'.

> **Keyword definitions**
>
> **Species** – A group of organisms that interbreed and are capable of producing fertile offspring.
>
> **Population** – A group of organisms of the same species living in the same area at the same time, and which are capable of interbreeding.
>
> **Habitat** – The environment in which a species normally lives.
>
> **Niche** – A species' share of a habitat and the resources in it. An organism's ecological niche depends not only on where it lives but also on what it does.

■ Niches

- A niche describes the particular set of abiotic and biotic conditions and resources to which an organism or population responds.
- The niche is a complete description of a species – where, when and how it lives.
- Two species cannot share the same niche.
- Species with overlapping niches will compete with each other (see below).

The **fundamental niche** can be described as where and how an organism *could* live, and the **realised niche** as where and how an organism *does* live (Figure 2.2).

> **Expert tip**
>
> Habitats can change over time as a result of migration.

> ### ■ QUICK CHECK QUESTIONS
>
> 1 Define the term *species*.
> 2 Distinguish between the terms *habitat* and *niche*.

Table 2.1 Comparing the fundamental and realised niche

Fundamental niche	Realised niche
Defines the physical conditions under which a species might live, in the absence of interactions with other species	Interactions such as competition, predation, disease and parasitism may restrict the environments in which a species can live; these more restricted conditions are the realised niche
May not be fully occupied, due to the presence of competing species, the absence of required positive species interactions, or dispersal limitations	Defined by the presence of competing species, the absence of required positive species interactions, or dispersal limitations

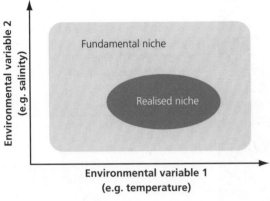

Figure 2.2 The fundamental and realised niche

> **Keyword definitions**
>
> **Fundamental niche** – The full range of conditions and resources in which a species could survive and reproduce.
>
> **Realised niche** – The actual conditions and resources in which a species exists due to biotic interactions.

Abiotic and biotic factors

Revised ☐

Abiotic factors include:

- **marine** – turbidity, salinity, pH, temperature, dissolved oxygen, wave action
- **freshwater** – turbidity, flow velocity, pH, temperature, dissolved oxygen
- **terrestrial** – temperature, light intensity, wind speed, particle size, slope, soil moisture, drainage, mineral content.

Biotic factors include the different species in an ecosystem (the community – see page 32) and the interactions between them.

> **Keyword definitions**
>
> **Abiotic factor** – A non-living, physical factor that can influence an organism or ecosystem – for example, temperature, sunlight, pH, salinity or precipitation.
>
> **Biotic factor** – A living part of an ecosystem (i.e. part of the community) that can influence an organism or ecosystem.

Population interactions

Revised ☐

Ecosystems contain many different species, with different interactions being shown between populations.

Competition between organisms can be:

- **intraspecific** – competition between members of the *same* species
- **interspecific** – competition between members of *different* species.

Intraspecific competition affects the **carrying capacity** of a population. Populations with high numbers per area (i.e. higher density) will experience more intraspecific competition and therefore have a lower carrying capacity due to the lower amount of food available. In interspecific competition, species with similar niches will compete for similar resources, lowering the carrying capacity of one or both of the species.

Parasitism is a relationship where one organism – the **parasite** – benefits at the expense of another – the **host** – from which it derives its food. The parasite lowers the carrying capacity of the host.

> **Keyword definitions**
>
> **Competition** – A common demand by two or more organisms for a limited supply of a resource such as food, water, light, space, mates and nesting sites.
>
> **Carrying capacity** – The maximum number of a species or 'load' that can be sustainably supported by a given environment.

Diseases are caused by pathogens, which include bacteria, viruses, **fungi** and single-celled animals (protozoa). A pathogen may reduce the carrying capacity of the organism it is infecting (Figure 2.3). Changes in the incidence of the pathogen, and therefore the disease, can also cause populations to increase and decrease around the carrying capacity.

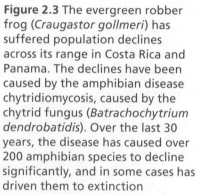

Figure 2.3 The evergreen robber frog (*Craugastor gollmeri*) has suffered population declines across its range in Costa Rica and Panama. The declines have been caused by the amphibian disease chytridiomycosis, caused by the chytrid fungus (*Batrachochytrium dendrobatidis*). Over the last 30 years, the disease has caused over 200 amphibian species to decline significantly, and in some cases has driven them to extinction

Expert tip

You need to know how each species in an interaction influences the population dynamics of the others, and the carrying capacity of the others' environment.

Mutualism (symbiosis) is an interaction in which both species derive benefit. Mutualism can increase the carrying capacity of both species in the relationship. An example of mutualism is coral, which is a mutualistic relationship between an animal polyp and photosynthetic algae (see page 48). Lichens are another example (Figure 2.4).

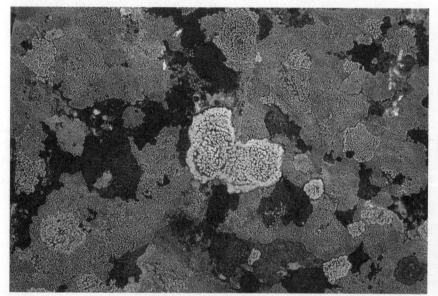

Figure 2.4 A lichens is a symbiotic relationship between a fungus and an alga

Figure 2.5 Pitcher plants feed on insects trapped in their highly adapted leaves (the pitchers)

Predation is when one animal (or sometimes a plant – see Figure 2.5) eats another animal. The number of prey is reduced by the predator, lowering the prey's carrying capacity. The carrying capacity of the predator is affected by the prey because the number of predators is reduced when prey become fewer (see predator–prey interactions, Chapter 1, page 13).

Herbivory is when an organism feeds on a plant. The carrying capacity of herbivores is affected by the quantity of plants they feed on. An area abundant in plant resources will have a higher carrying capacity for herbivores than an area that has less plant material.

The interactions between the organisms – such as predation, herbivory, parasitism, mutualism, disease and competition – are all biotic factors.

Limiting factors and carrying capacity

Revised

Limiting factors restrict the growth of a population or prevent it from increasing further.

- Limiting factors in plants include light, nutrients, water, carbon dioxide and temperature.
- Limiting factors in animals include space, food, mates, nesting sites and water.

When a graph of population growth is plotted against time, an **S-population curve** is generally produced (Figure 2.6).

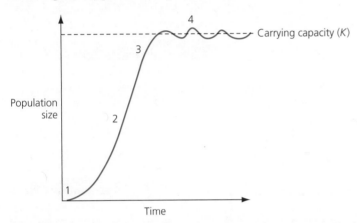

Figure 2.6 Population growth curve controlled by limiting factors

- There is slow initial growth when the population is small (point 1 in Figure 2.6).
- With low or reduced limiting factors the population expands **exponentially** into the habitat (point 2).
- As a population grows then there will be increased competition between the individuals of that population for the same limiting factors, i.e. resources.
- This results in a lower rate of population increase (point 3).
- Population reaches its **carrying capacity** (see page 29), fluctuating around a set point determined by the **limiting factors** (point 4).
- Changes in limiting factors cause the population size to increase and decrease.
- Increases and decreases around the carrying capacity are controlled by **negative feedback** mechanisms.

■ **QUICK CHECK QUESTIONS**

3   Define the term *carrying capacity*.
4   Explain why populations fluctuate around a set point.

S- and J-population curves

Revised

■ S-population curve

The S-population curve (Figure 2.6), showing the establishment of a population following introduction into a new environment, has four distinct phases, with each phase given a specific name:

- **Lag phase**, where population numbers are low, leading to low birth rates (point 1, Figure 2.6).
- **Exponential growth phase**, where limiting factors are not restricting the growth of a population (point 2).
- **Transitional phase**, as limiting factors begin to affect the population and restrict its growth (point 3).

■ **Plateau phase**, where limiting factors restrict the population to its carrying capacity (point 4). Changes in limiting factors, predation, disease and abiotic factors cause populations to increase and decrease around the carrying capacity (*K*).

■ J-population curves

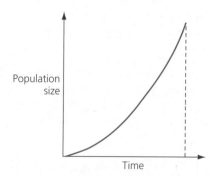

Figure 2.7 J-population growth curve

J-population curves show only exponential growth. Growth is initially slow but becomes increasingly rapid, and does not slow down as population increases (Figure 2.7).

■ The population is not controlled by limiting factors in the exponential growth phase.
■ After reaching its peak value, the population will suddenly decrease (dotted line on Figure 2.7).

Populations showing J-shaped curves are controlled by abiotic but not biotic factors. Abiotic factors cause the sudden decrease in the population (the **population crash**).

> **Expert tip**
>
> S- and J-population curves describe a generalised response of populations to a particular set of conditions (abiotic and biotic factors).

> ■ **QUICK CHECK QUESTIONS**
>
> 5 Outline the stages of an S-population growth curve.
> 6 Explain the differences between S- and J-population growth curves.

> **Expert tip**
>
> You need to be able to explain population growth curves in terms of numbers and rates.

> **EXAM PRACTICE**
>
> 1 Discuss how negative feedback processes control populations of a named parasite.   [3]

2.2 Communities and ecosystems

Revised ▢

> **SIGNIFICANT IDEAS**
>
> • The interactions of species with their environment result in energy and nutrient flows.
> • Photosynthesis and respiration play a significant role in the flow of energy in communities.
> • The feeding relationships of species in a system can be modelled using food chains, food webs and ecological pyramids.

Communities and ecosystems

Revised ▢

The Earth's **biosphere** is a narrow zone, a few kilometres in thickness. It extends from the upper part of the atmosphere (where birds, insects and windblown pollen may be found) down to the deepest part of the Earth's crust to which living organisms venture. The biosphere is made up of **ecosystems** (Figure 2.8), which have biotic and abiotic components. The biotic component is the **community** (Figure 2.9).

Figure 2.8 All ecosystems, such as tropical rainforest, include both living (e.g. animals, plants and fungi) and non-living components

> **Expert tip**
>
> Make sure you know and understand the difference between the terms *biotic* and *abiotic*: biotic refers to living components of the ecosystem and abiotic to non-living components.

> **Keyword definitions**
>
> **Biosphere** – That part of the Earth inhabited by organisms.
>
> **Ecosystem** – A community and the physical environment with which it interacts.
>
> **Community** – A group of populations living and interacting with each other in a common habitat.

> ■ **QUICK CHECK QUESTIONS**
>
> 7 Define the term *ecosystem*.
> 8 Identify *three* biotic and *three* abiotic components of an ecosystem near to where you live.

Figure 2.9 A community of different animals in an African savannah. The savannah is the habitat in which the animals live

The interactions of species within their environment results in energy and nutrient flows.

Photosynthesis and respiration

Revised ☐

The flow of energy through ecosystems is the result of photosynthesis and respiration.

Photosynthesis (Figures 2.10 and 2.11) is the transformation of light energy into the chemical energy of organic matter:

- Carbon dioxide, water, chlorophyll and certain visible wavelengths of light (e.g. red and blue) are used to produce organic matter (glucose) and oxygen.
- The process is controlled by enzymes – warmer conditions increase the rate of photosynthesis up to an optimum temperature.

$$\text{carbon dioxide } + \text{ water } \xrightarrow[\text{chlorophyll}]{\text{light}} \text{ glucose } + \text{ oxygen}$$

Figure 2.10 The word equation for photosynthesis

> **Expert tip**
>
> All organisms respire – bacteria, protoctists (e.g. algae), fungi, plants and animals.

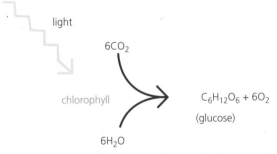

Figure 2.11 The chemical reactions of photosynthesis

Respiration (Figure 2.12) is the breakdown of glucose using oxygen, releasing carbon dioxide, water and energy:

- Stored chemical energy is transformed into kinetic energy and heat.
- Respiration carried out using oxygen is called aerobic respiration.
- Respiration carried out without oxygen is called anaerobic respiration, where carbon dioxide and other waste products are formed.
- Energy is released in a form available for use by living organisms, but is ultimately transformed into heat (second law of thermodynamics).

$$\text{glucose } + \text{ oxygen } \longrightarrow \text{ carbon dioxide } + \text{ water } (+ \text{ ENERGY})$$

Figure 2.12 The word equation for respiration

> **Expert tip**
>
> Although respiration involves the release of energy, energy is not generally included in the word equation.

Table 2.2 Input, outputs and transformations of photosynthesis and respiration

Process	Inputs	Outputs	Transformations
Photosynthesis	Sunlight energy Carbon dioxide Water	Glucose Oxygen	Light energy into stored chemical energy
Respiration	Glucose Oxygen	Carbon dioxide Water Energy	Stored chemical energy into kinetic energy and heat

Expert tip

You need to be able to construct system diagrams representing photosynthesis and respiration.

Food chains and food webs

Revised

Feeding relationships in ecosystems can be modelled using food chains and food webs.

Figure 2.13 shows a typical food chain.

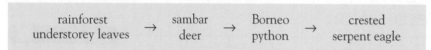

rainforest understorey leaves $\rightarrow$ sambar deer $\rightarrow$ Borneo python $\rightarrow$ crested serpent eagle

Figure 2.13 A rainforest food chain from Borneo

The rainforest food chain in Figure 2.13 contains the following trophic levels:

- **producer** – rainforest understorey leaves
- **primary consumer** – sambar deer (**herbivore**)
- **secondary consumer** – Borneo python (**carnivore**)
- **tertiary consumer** – crested serpent eagle (**top carnivore**).

Ecosystems contain many interconnected food chains that form **food webs** (Figure 2.14). These show the complex feeding relationships that exist among species.

Keyword definition

Trophic level – The position that an organism occupies in a food chain, or a group of organisms in a community that occupy the same position in food chains.

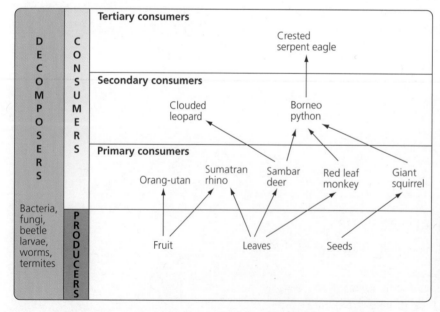

Figure 2.14 A rainforest food web from Borneo, showing trophic levels; decomposers feed at each trophic level

You do not need to draw pictures of the plants and animals in food chains.

Expert tip

If asked to draw a food chain, you should use specific organisms (e.g. 'Borneo python' rather than 'snake') from a specific ecosystem (e.g. 'Borneo rainforest' rather than 'forest').

Common mistake

Arrows in food chains represent energy flow and always run from left to right, towards the consumers, not towards the organisms being eaten

Common mistake

If you are asked to draw a food chain, do not draw a food web, or vice versa. A food chain is linear, showing energy flow through an ecosystem. A food web shows the complex interactions between different food chains.

Pyramids of numbers, biomass and productivity

Pyramids are graphical models. They show quantitative differences between the trophic levels in an ecosystem.

■ Pyramids of numbers

A **pyramid of numbers** represents the number of organisms (producers and consumers) coexisting in an ecosystem (Figure 2.15).

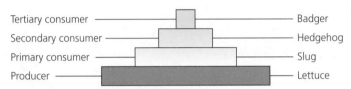

Tertiary consumer ——————— Badger
Secondary consumer ——————— Hedgehog
Primary consumer ——————— Slug
Producer ——————— Lettuce

Figure 2.15 A pyramid of numbers

> **Expert tip**
>
> You need to be able to construct models of feeding relationships – such as food chains, food webs and ecological pyramids – from given data.

Table 2.3 The strengths and weaknesses of pyramids of numbers

Strengths	Weaknesses
• A simple method of giving an overview of community structure.	• They do not take into account the size of organisms (page 36).
• Good for comparing changes in a number of individuals over time.	• Numbers can be too great to represent accurately.
	• Some animals feed at more than one trophic level (**omnivores**) and are therefore difficult to place.

■ Plotting a pyramid of numbers

Quantitative data for each trophic level are drawn to scale as horizontal bars, arranged symmetrically around a central axis.

Worked example

Construct a pyramid of numbers from the following data.

- Draw two axes on graph paper – the vertical axis should be located centrally on the paper (Figure 2.16 – axes are shown as dotted lines).
- Data are plotted symmetrically around the vertical axis, for example there are 16 lettuces, and so the vertical bar is drawn with eight units to the left and eight to the right of the axis.
- The height of the bars is arbitrary, but each bar should be the same height.
- Label each trophic level.

Species	Number of individuals
Lettuce	16
Slug	10
Hedgehog	6
Badger	4

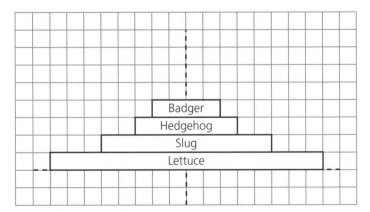

Figure 2.16 Plotting a pyramid of numbers

■ Pyramids of biomass

A **pyramid of biomass** shows the biological mass at each trophic level. Each trophic level is measured in grams of biomass per square metre ($g\,m^{-2}$) or kilograms per square metre ($kg\,m^{-2}$). Biomass can also be measured in units of energy (e.g. $J\,m^{-2}$).

- Because energy decreases along food chains (second law of thermodynamics) biomass decreases along food chains, so pyramids become narrower towards higher trophic levels.
- Biomass is measured as dry weight (biological mass minus water).
- Both pyramids of numbers and pyramids of biomass represent storages.

Pyramids of numbers and biomass can be inverted, i.e. narrower at the base than at the next trophic level:

- If producers, such as an oak tree, are relatively large in size and so few in number, there will be fewer producers than primary consumers in a food chain (Figure 2.17).

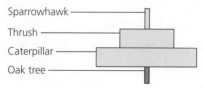

Figure 2.17 Pyramids of numbers do not take into account the size of the organisms, and can therefore be inverted

- Data for numbers and biomass pyramids are taken at a point in time. The biomass of the producers may be less than the consumers that feed on them – this also leads to the pyramid of biomass being inverted (Figure 2.18).

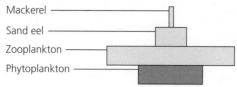

Figure 2.18 A pyramid of biomass for a north Atlantic food chain. Changes in feeding patterns and seasonal variations can lead to pyramids of biomass being inverted

■ Pyramids of productivity

If data are taken over a full year, the total biomass produced at the producer level will always be greater than the primary consumer level, and so on through the food chain, following the second law of thermodynamics. Such data are shown as **pyramids of productivity**. Pyramids of productivity refer to the flow of energy through trophic levels and always show a decrease in energy along the food chain (Figure 2.19).

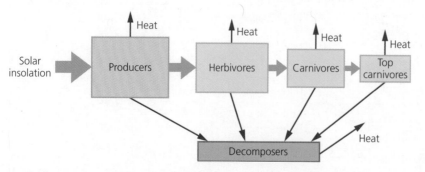

> **Common mistake**
>
> Units are often not included when describing pyramids of biomass – these must not be forgotten (e.g. grams of biomass per square metre, $g\,m^{-2}$).

> **Expert tip**
>
> 'Pyramids of biomass' refers to a standing crop (a fixed point in time) and 'pyramids of productivity' refer to the rate of flow of biomass or energy.

> **Expert tip**
>
> You need to be able to explain the relevance of the laws of thermodynamics to the flow of energy through ecosystems.

Figure 2.19 Diagram showing the transfers and transformations of energy as it flows through an ecosystem. Arrows showing flows of energy vary in width, proportional to the amount of energy being transferred

> **Expert tip**
>
> Storages of energy are shown by boxes in energy-flow diagrams (representing the various trophic levels), and the flows of energy or productivity are shown as arrows, sometimes of varying widths to indicate differences between the magnitudes of the different energy flows.

> **Expert tip**
>
> You need to be able to analyse quantitative models of flows of energy and matter.

Pyramids of productivity are the only pyramids that are always pyramid shaped:

- Pyramids of numbers and biomass show the storage in the food chain at a given time, whereas pyramids of productivity show the rate at which those storages are being generated.
- Productivity is defined by the amount of new biomass created per unit area per unit time. It is measured in units of flow (e.g. $gm^{-2}yr^{-1}$ or $Jm^{-2}yr^{-1}$).

> **Expert tip**
>
> Biomass, measured in units of mass or energy (e.g. gm^{-2} or Jm^{-2}), should be distinguished from productivity, measured in units of flow (e.g. $gm^{-2}yr^{-1}$ or $Jm^{-2}yr^{-1}$).

> **Expert tip**
>
> You need to be able to construct a quantitative model of the flows of energy or matter for given data.

Pyramid structure and ecosystem functioning

Revised ☐

Food chains and their respective pyramids can be changed by human activities:

- Because energy is lost through food chains, top carnivores tend to be few in number. Any reduction in the number of organisms lower down the food chain can therefore lead to carnivores being put at risk.
- Crop farming increases producers (base of pyramid) and decreases higher trophic levels.
- Livestock farming increases primary consumers and decreases secondary and tertiary consumers.
- Hunting removes top carnivores.
- Deforestation reduces the producer bar on biomass pyramids.
- The use of non-biodegradable toxins (such as DDT or mercury) can lead to reduction in the length of food chains:
  - ☐ The toxin accumulates in the body fat of consumers.
  - ☐ The toxin becomes increasingly concentrated as the consumers at each trophic level become fewer.
  - ☐ Organisms higher up the food chain live longer and so have more time to accumulate the toxin.
  - ☐ Top carnivores are at risk from poisoning from the toxin.

> **■ QUICK CHECK QUESTIONS**
>
> 9 Explain the differences between pyramids of numbers, biomass and productivity.
> 10 Explain why pyramids of numbers and biomass can sometimes be inverted.
> 11 How can human activities affect pyramid structure? Give *three* examples.

> **Expert tip**
>
> You need to be able to explain the impact of a persistent or non-biodegradable pollutant in an ecosystem.

> **EXAM PRACTICE**
>
> 2 Using examples, distinguish between a food chain and a food web.   [5]

> **Keyword definitions**
>
> **Bioaccumulation** – The build-up of persistent or non-biodegradable pollutants within an organism or trophic level because they cannot be broken down.
>
> **Biomagnification** – The increase in concentration of persistent or non-biodegradable pollutants along a food chain.

> **Expert tip**
>
> Toxins such as DDT and mercury accumulate along food chains due to the decrease of biomass and energy.

2.3 Flows of energy and matter

SIGNIFICANT IDEAS
- Ecosystems are linked together by energy and matter flows.
- The Sun's energy drives these flows, and humans are impacting the flows of energy and matter both locally and globally.

Pathway of energy entering the atmosphere

As solar radiation (**insolation**) enters the Earth's atmosphere, not all energy reaches the surface and is lost through reflection and absorption. This lost energy is therefore unavailable for ecosystems. Energy is:

- absorbed by molecules and dust in the atmosphere and by clouds
- reflected by scatter, clouds and the ground.

Pathways of radiation through the atmosphere involve a loss of radiation, as shown in Figure 2.20.

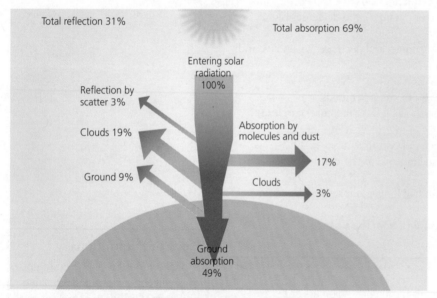

Figure 2.20 Pathway of energy entering the atmosphere

Expert tip

Pathways of energy through an ecosystem include:
- conversion of light energy to chemical energy
- transfer of chemical energy from one trophic level to another with varying efficiencies
- overall conversion of ultraviolet and visible light to heat energy by an ecosystem
- re-radiation of heat energy to the atmosphere.

Transfer and transformation of energy

Energy enters the ecosystem as sunlight energy, is transformed into chemical energy/biomass, is then transferred between trophic levels by consumers and ultimately leaves the ecosystem as heat energy.

- Very little of the available sunlight energy is used to make new biomass.
- Producers are inefficient at converting sunlight energy into stored chemical energy (Figure 2.21).

In a food chain there is a loss of chemical energy from one trophic level to another. **Ecological efficiency** is the percentage of energy transferred from one trophic level to the next:

$$\text{ecological efficiency} = \frac{\text{energy used for growth (new biomass)}}{\text{energy supplied}} \times 100$$

- Efficiencies of transfer are low and account for energy loss. They vary from 5% to 20%, with an average of 10%.
- Energy is lost through movement, inedible parts (e.g. bone, teeth, fur), faeces, and ultimately as heat through the inefficient energy conversions of respiration (second law of thermodynamics).
- Overall there is the conversion of light to heat energy by an ecosystem.
- Energy is converted from one form to another but cannot be created or destroyed (first law of thermodynamics).
- Inputs of the system as a whole, and of any individual trophic level, equal the outputs.

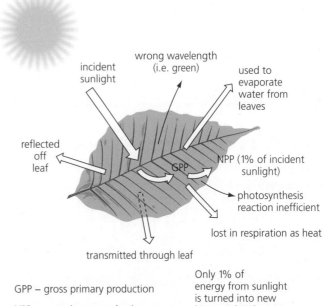

GPP – gross primary production

NPP – net primary production

Only 1% of energy from sunlight is turned into new biomass (NPP)

Figure 2.21 The conversion of sunlight energy into new biomass is inefficient.

Expert tip

You need to be able to analyse the efficiency of energy transfers through a system.

Worked example

100 units of energy are contained in the biomass of the caterpillar.

The shrew eats the caterpillar. 10 units of energy go into forming new biomass, 60 units are lost as gas, faeces and waste, and 30 through respiration and heat loss.

90 units of energy are therefore lost from the food chain, and 10 units remain in the shrew.

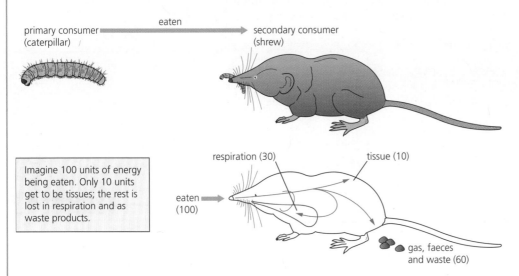

Imagine 100 units of energy being eaten. Only 10 units get to be tissues; the rest is lost in respiration and as waste products.

Figure 2.22 Calculating the ecological efficiency for energy flow from a caterpillar to a shrew

$$\text{ecological efficiency} = \frac{\text{energy used for new growth in the shrew}}{\text{energy supplied to the shrew when it eats the caterpillar}} \times 100 = \frac{10}{100} \times 100 = 10\%$$

■ **QUICK CHECK QUESTIONS**

12 Explain the differences between photosynthesis and respiration. Use the terms *transfer* and *transformation* in your answer.

13 Outline the transfers and transformations of energy as it flows through an ecosystem.

14 Explain the term *ecological efficiency*.

Expert tip

You need to be able to explain the transfer and transformation of energy as it flows through an ecosystem.

Expert tip

Make sure you understand the difference between storages and flows of energy:

- Storages of energy are shown as boxes, representing the various trophic levels, and measured as the amount of energy or biomass per unit area.
- Flows of energy or productivity are given as rates (i.e. inclusive of time), for example $J\,m^{-2}\,day^{-1}$.

Expert tip

You need to be able to analyse the efficiency of energy transfers through a system.

Gross productivity, net productivity, primary productivity and secondary productivity

Revised ☐

Keyword definitions

Gross productivity (GP) – The total gain in energy or biomass per unit area per unit time, which could be through photosynthesis in primary producers or absorption in consumers.

Net productivity (NP) – The gain in energy or biomass per unit area per unit time remaining after allowing for respiratory losses (R).

Primary productivity – The gain by producers in energy or biomass per unit area per unit time. This term could refer to either gross or net primary productivity.

Secondary productivity – The biomass gained by consumers (heterotrophic organisms), through feeding and absorption, measured in units of mass or energy per unit area per unit time.

Expert tip

Productivity is production per unit time. You must include units when defining it.

Gross primary productivity and net primary productivity

Revised ☐

Keyword definitions

Gross primary productivity (GPP) – The total gain in energy or biomass per unit area per unit time fixed by photosynthesis in green plants.

Net primary productivity (NPP) – The gain by producers in energy or biomass per unit area per unit time remaining after allowing for respiratory losses (R). This is potentially available to consumers in an ecosystem.

$$NPP = GPP - R$$
(where R = respiratory loss)

Figure 2.23 Equation for net primary productivity

NPP is the rate at which plants accumulate new biomass. It represents the actual store of energy contained in potential food for consumers. NPP is easier to calculate than **GPP** as biomass is simpler to measure than the amount of energy fixed into glucose.

Expert tip

You need to be able to calculate the values of both GPP and NPP from given data.

Gross secondary productivity and net secondary productivity

Revised

- Much of the biomass eaten by consumers is absorbed (e.g. through the guts of animals) and converted into new biomass within cells.
- Consumers do not use all the biomass they eat.
- Some energy passes out in faeces and excretion.
- Only the biomass remaining can be used by the consumer **(GSP)**.
- Some of the biomass absorbed by animals is used in respiration.
- The energy released is used to support life processes.
- The remaining energy is available to form new biomass **(NSP)**.
- This new biomass is then available to the next trophic level.

GSP = food eaten – faecal loss
NSP = GSP – R
(where R = respiratory loss)

Figure 2.24 Equations for GSP and NSP

Keyword definitions

Gross secondary productivity (GSP) – The total gain by consumers in energy or biomass per unit area per unit time through absorption.

Net secondary productivity (NSP) – The gain by consumers in energy or biomass per unit area per unit time remaining after allowing for respiratory losses (R).

■ QUICK CHECK QUESTIONS

15 Distinguish between primary productivity and secondary productivity.
16 Outline how you would calculate net primary productivity.
17 Explain the differences between gross secondary productivity and net secondary productivity.

Expert tip

You need to be able to calculate the values of both GSP and NSP from given data.

Maximum sustainable yield

Revised

The annual **sustainable yield** for a natural resource such as a forest is the annual gain in biomass or energy through growth and recruitment.

Maximum sustainable yield is the highest rate of harvesting that does not lead to a reduction in the original natural capital. It is equivalent to the net primary or net secondary productivity of a system. Net productivity is measured as the amount of energy stored as new biomass per year – any removal of biomass at a rate greater than this would indicate that NPP or NSP would not be able to replace the biomass that had been extracted.

Any harvesting that occurs above these levels is therefore unsustainable and will lead to a reduction in the natural capital.

Keyword definition

Sustainable yield – When a natural resource can be harvested at a rate equal to or less than its natural productivity so that the natural capital is not diminished.

Expert tip

You need to understand the link between sustainable yields and productivity.

Transfer and transformation of materials within an ecosystem

Materials such as carbon, nitrogen and water are cycled within an ecosystem. These cycles involve transfer and transformation processes including the conversion of organic and inorganic storages (Figure 2.25). The cycles involve producers, consumers and decomposers.

Producers make their own food (glucose) and convert (fix) inorganic molecules into organic molecules (Figure 2.25). Plants, algae and some bacteria are producers:

- **Photoautotrophs** (e.g. all plants) convert sunlight energy into chemical energy.
- **Chemoautotrophs** (e.g. nitrifying bacteria) use chemical energy from oxidation reactions to create glucose.

Producers support all ecosystems through constant input of energy and new biomass.

Consumers do not contain photosynthetic pigments (e.g. chlorophyll) and so cannot make their own food. They must obtain the energy, minerals and nutrients they need by eating other organisms. They are also known as **heterotrophs**. They pass energy and biomass through a food chain from producers through to top carnivores.

Decomposers obtain their food from the breakdown of dead organic matter. They include bacteria and fungi. Decomposers release nutrients ready for absorption by producers (Figure 2.25). They feed at each trophic level of a food chain and are essential for recycling matter, including elements such as carbon and nitrogen, in ecosystems.

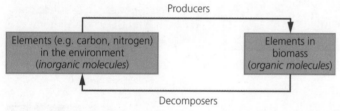

Figure 2.25 The role of producers and decomposers in ecosystems

---

■ **QUICK CHECK QUESTIONS**

18 Explain the differences between producers and consumers.

19 Explain the role of producers and decomposers in ecosystems.

The cycles shown in Figures 2.26 and 2.27 illustrate the different ways cycles can be represented.

■ Carbon cycle

Storages in the carbon cycle (Figure 2.26) include:

- organic storage – organisms and forests
- inorganic storage – atmosphere, soil, fossil fuels and oceans.

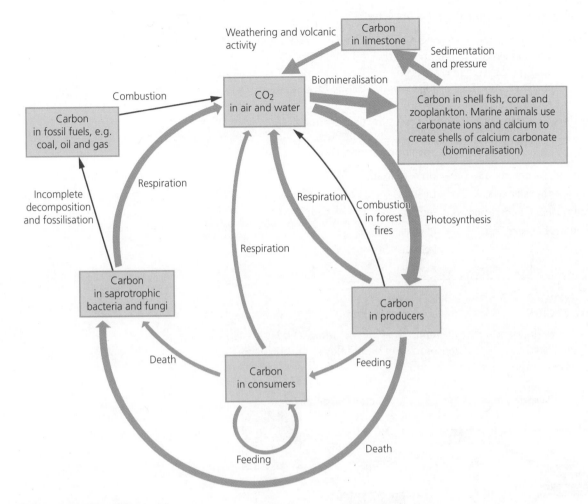

Figure 2.26 The carbon cycle

Table 2.4 Transfers and transformation processes in the carbon cycle

Transfers	Transformations
Feeding on plant material by herbivores	Photosynthesis (CO_2 into glucose)
Feeding on herbivores by carnivores	Respiration (organic matter into CO_2)
Feeding on dead organisms by decomposers	Combustion (organic matter into CO_2)
CO_2 from atmosphere dissolving in rainwater	Incomplete decomposition and fossilisation
CO_2 from atmosphere dissolving in oceans	

■ Nitrogen cycle

Storages in the nitrogen cycle (Figure 2.27) include:

- ■ organic storage – organisms
- ■ inorganic storage – soil, fossil fuels, atmosphere and water bodies.

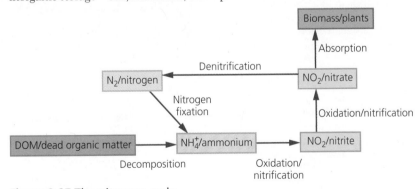

Figure 2.27 The nitrogen cycle

The nitrogen cycle involves four different types of **bacteria**:

- **Nitrogen fixing** – nitrogen from atmosphere converted into ammonium ions.
- **Nitrifying** – ammonium ions converted into nitrite and then nitrate.
- **Denitrifying** – nitrates converted into nitrogen.
- **Decomposers** – break down organic nitrogen into ammonia (**deamination**).

Table 2.5 Transfers and transformation processes in the nitrogen cycle

Transfers	Transformations
Feeding on plant material by herbivores	Nitrogen fixation
Feeding on herbivores by carnivores	Lightning
Feeding on dead organisms by decomposers	Nitrification
Absorption of nitrates by plants	Denitrification
	Deamination
	Assimilation

■ Nutrient cycles

Nutrient cycles can be drawn as simple diagrams, showing the storages and flows of nutrients. Figure 2.28 shows the differences in storages and flows of nutrients in two contrasting ecosystems.

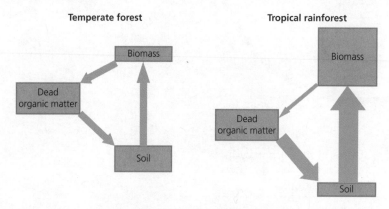

Figure 2.28 Diagrams showing the major nutrient flows and storages in two different ecosystems. The size of the boxes and width of the arrows are proportional to the size of the storages and flows they represent

Human impacts on energy flows and matter cycles

■ Energy flows

Industrialisation has led to the increased use of fossil fuels.

- Energy trapped by plants millions of years ago is being released
- The amount of energy available to humans has increased enormously, enabling industry and agriculture to be powered.
- Population growth, through increased food output, has increased rapidly.
- This change in the Earth's energy budget has led to many environmental issues – habitat destruction, climate change and the reduction of non-renewable resources, for example.
- Increased carbon dioxide levels and the corresponding increase in temperatures have led to the reduction in Arctic land and sea ice, leading to a potential tipping point (Chapter 1, pages 14–15).

Expert tip

You need to be able to discuss human impacts on energy flows, and on the carbon and nitrogen cycles.

■ Matter cycles

■ **Carbon cycle**

Human activity has many impacts on the carbon cycle.

■ Urbanisation

- Increases the need for energy and therefore increased use of fossil fuels (increased combustion – see below).
- Leads to decreased land covered by vegetation, reducing photosynthesis.
- Increases food requirements, leading to increased land use for agriculture (see below).
- Increased transport of food leads to greater energy requirements and increased fossil fuel use (see below).

■ Deforestation

- Carbon storages are reduced as trees are removed.
- Soil erosion is made worse, leading to the flow of carbon stored in soil into the rivers.
- A reduction in photosynthesis and therefore less carbon dioxide removed from the atmosphere.
- Increase respiration from decomposers feeding on decaying forest residues.

■ Agriculture

- Increased land use for agriculture rather than native ecosystems, altering the nature of carbon storage.
- The carbon storage present in crops is transported to new locations, altering the carbon cycle on a local and global scale.

■ Fossil fuel use

- Involves the direct burning of a carbon store locked up for millions of years in geological deposits.
- Mining and burning of fossil fuels reduces the storages of these non-renewable sources of energy and increases the storage of carbon in the atmosphere.
- Increased release of greenhouse gases, for example carbon dioxide, and therefore warmer average global temperatures.
- Increased carbon dioxide levels in the atmosphere can lead to increased vegetation growth, due to increased carbon dioxide available for photosynthesis, altering the carbon cycle.

■ **Nitrogen cycle**

Human impacts on the nitrogen cycle include the following.

■ Agriculture

- Nitrogen fixation via industrial techniques, such as the Haber process, has significantly increased the amount of global nitrogen fixation, leading to increased amounts of useable nitrogen in the form of fertilisers.
- Application of nitrate fertiliser increases the amount of biologically available nitrogen in an ecosystem.
- Nitrate fertiliser, used to increase crop yield, runs off or leaches into bodies of waters, such as rivers and lakes, causing eutrophication and disruption to ecosystems (see pages 100–102).
- Eutrophication leads to low oxygen levels in aquatic ecosystems, changing food-web structure, resulting in habitat degradation.
- The addition of nitrogen can lead to changes in biodiversity and species composition that may lead to changes in overall ecosystem function.
- Nitrogen in the biomass of crops is transferred from fields in one area to markets in other areas. These processes remove nitrogen from the cycle in

one location and add it to another cycle at a different location. This alters the nitrogen cycle and can cause disruption to ecosystems.

- Waterlogged soils on agricultural land leads to an increase in denitrifying bacteria, increasing the rate at which nitrogen gas is returned to the atmosphere.

■ Deforestation

- Trees store nitrogen in the form of amino acids and protein. When trees are removed, this storage is lost.
- Logging increases the amount of atmospheric nitrogen and decreases land-based storages.

■ Fossil fuel use

- Fossil fuels such as coal contain nitrogen. When fossil fuels are burned, they release nitrogen oxides into the atmosphere, which contribute to the formation of smog (see Chapter 6, pages 123–125) and acid rain (pages 126–128).
- Burning fossil fuels releases nitrogen from storage in geological deposits and increases storages in the atmosphere, land and sea.
- Increases the amount of biologically available nitrogen in an ecosystem.

■ Human population growth

- Increases in human population have led to increased food needs and production. To support the increase in food needed to support the growing population, fertilisers have been used to increase crop yield (see above).
- Increased sewage output leads to increased quantities of ammonium and nitrates in rivers, lakes and the sea.

EXAM PRACTICE

3 a Suggest how the differences in the sizes of comparable storages of the two ecosystems shown in Figure 2.28 (page 44) can be explained in terms of their different climates. [7]

  b Draw a labelled flow diagram showing the flows and storages of inorganic nitrogen that normally occur within soil. Show on your diagram how these flows provide a link between the storages of dead organic matter and biomass. [3]

4 State *two* transfers and *two* transformations in the carbon cycle. [2]

5 Using a diagram, show how carbon is cycled through an ecosystem. [4]

6 Explain the relationship between climate and net primary productivity in *two* contrasting biomes you have studied. [7]

7 State the *two* factors that would need to be measured in order to estimate the gross productivity of an animal population in $g\,m^{-2}\,yr^{-1}$. [2]

2.4 Biomes, zonation and succession

Revised

SIGNIFICANT IDEAS

- Climate determines the type of biome in a given area, although individual ecosystems may vary due to many local abiotic and biotic factors.
- Succession leads to climax communities that may vary due to random events and interactions over time. This leads to a pattern of alternative stable states for a given ecosystem.
- Ecosystem stability, succession and biodiversity are intrinsically linked.

Biomes

Revised ☐

Biomes are collections of ecosystems sharing similar climatic conditions that can be grouped into five major classes: aquatic, forest, grassland, desert and tundra. Each of these classes has characteristic limiting factors, productivity and biodiversity. Biome distribution depends on levels of insolation (sunlight), temperature and precipitation (rainfall). The model in Figure 2.29 shows the relationship between temperature, precipitation and biome. It shows the likely stable biomes that are found under specific climatic conditions.

Plant growth strongly influences the distribution of different biomes.

Temperature and rainfall are two of the main limiting factors that affect plant growth, and so these abiotic factors can be used to model and predict the geographical distribution of different ecosystems around the planet.

> **Keyword definition**
>
> **Biome** – A collection of ecosystems sharing similar climatic conditions – for example, tundra, tropical rainforest and desert.

> **Expert tip**
>
> You need to be able to explain the distribution, structure, biodiversity and relative productivity of contrasting biomes. Climate should be explained in terms of temperature, precipitation and insolation only.

> **Expert tip**
>
> You should study at least four contrasting pairs of biomes, such as temperate forests and tropical rainforests, tundra and deserts, tropical coral reefs and hydrothermal vents, or temperate bogs and tropical mangrove forests. You should focus on their relative limiting factors, productivity and biodiversity.

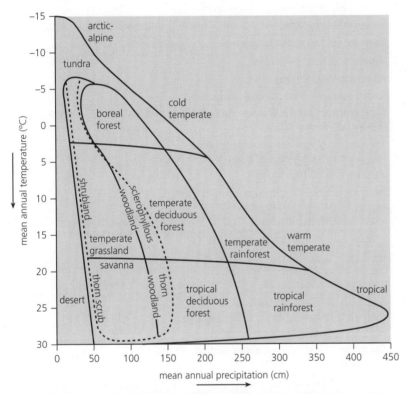

Figure 2.29 The distribution of major terrestrial biomes with respect to mean annual precipitation and temperature. The dashed line indicates biomes where factors other than rainfall and temperature strongly influence ecosystem structure, such as soil type, the occurrence of fire, animal grazing and seasonal drought

Table 2.6 Biome distribution, structure and productivity

Biome	Distribution/climate	Structure	Relative productivity
Temperate forest	Found between 40° and 60° N of equator Rainfall sufficient to establish forest (500–1500 mm yr⁻¹) rather than grassland Temperatures and light intensity vary with season	May contain seasonal (deciduous) trees, evergreen (e.g. coniferous trees) or both; evergreen trees have thicker leaves or needles to protect against the cold; deciduous trees lose their leaves in winter Less complex structure than rainforest, often dominated by one species Some layering of forest, although the tallest trees grow no more than around 30 m The lower and less dense canopy than rainforest means that more light reaches the forest floor – growth of rich shrub layer (e.g. brambles, grasses, bracken and ferns)	High: mean NPP = 1.20 kg m⁻² yr⁻¹

continued ...

Biome	Distribution/climate	Structure	Relative productivity
Tropical rainforest	Found between the Tropics of Cancer and Capricorn (23.5° N and S of equator) High rainfall (over 2500 mm yr^{-1}), sunlight and temperature No seasons, so consistent light and temperature	Has a complex structure with a number of layers from ground level to canopy Has emergent trees up to 50 m and lower layers of shrubs and vines Dense canopy means that only 1% of sunlight may reach the floor; shrub layer may be sparse with most productivity in the canopy Soils are thin and nutrient poor	Very high: mean net primary productivity (NPP) = 2.20 kg m^{-2} yr^{-1}
Desert	In bands at latitudes of approximately 15–30° N and S of equator Low rainfall (under 250 mm yr^{-1}); high sunlight; very hot in daytime and cold at night	Vegetation is scarce, with an absence of tall trees Many xerophytic plants (i.e. adapted to dry or desert conditions) such as cacti Soil has low water-holding capacity and low fertility Soil erodes easily in the wind Animals are adapted to desert conditions	Very low: mean NPP = 0.003 kg m^{-2} yr^{-1}
Tundra	High latitudes, adjacent to ice margins Low temperatures; low precipitation; seasonal sunlight and short day length	Has a simple structure Vegetation is low scrub and grasses Vegetation forms a single layer An absence of tall trees Frozen permafrost and soil limit productivity	Low: mean NPP = 0.14 kg m^{-2} yr^{-1}
Mangrove	Found in tropical coastal areas Average rainfall is generally over 1500 mm yr^{-1} and insolation is constant throughout year	Mangrove trees grow in a saline and oxygen-deficient environment The harsh environment limits productivity compared with other tropical ecosystems, although it is still high Aerial roots allow mangroves to absorb oxygen directly from the air and to survive when the forest floods with salt water Provide food, habitats and nursery-sites for many aquatic species	High: mean NPP = 1.2 kg m^{-2} yr^{-1}
Temperate bogs	Occur in areas where soil is acidic and nutrient-poor Found between 40° and 60° N of equator Rainfall 500–1500 mm yr^{-1} Temperatures and light intensity vary with season	Dominated by sphagnum moss, which can survive in the acidic conditions Acidic and waterlogged conditions do not allow establishment of forest so biodiversity is poor, although specialised species exist Around 1% of invertebrate species are bog specialists	Low: NPP = 0.4–0.8 kg m^{-2} yr^{-1}
Coral reef	Found between the Tropics of Cancer and Capricorn (23.5° N and S of equator) Seas are warm and there is strong sunlight throughout the year	Productivity of the symbiotic algae that live within the coral polyps is high, leading to a complex three-dimensional structure and high biodiversity	High: mean NPP = 2.0 kg m^{-2} yr^{-1} Areas of high productivity coincide with optimum conditions for growth
Hydrothermal vents	Found in volcanically active areas along tectonic plate margins, where cold seawater enters the ocean crust and comes into contact with hot rock below	Food chains are supported by chemosynthetic bacteria, in contrast to the food chains of coral reefs (supported by photosynthetic organisms – see above) Diversity is much lower than that found in coral reefs due to simpler ecosystem and fewer niches, although the productivity of the bacteria supports diverse communities of very specialised organisms not found elsewhere on the planet	Among the most productive ecosystems in the ocean, in contrast to the low productivity of surrounding areas NPP much lower than coral reef Areas of high productivity at vents coincide with harsh or stressful environmental conditions

■ QUICK CHECK QUESTIONS

20 State *two* factors that influence productivity
21 Explain how abiotic factors determine the distribution of tropical rainforest.
22 Compare the structure of tundra and a named local biome.

Expert tip

You need to be able to explain the distributions, structure, biodiversity and relative productivity of contrasting biomes.

Expert tip

You need to be able to analyse data for a range of biomes.

The tricellular model of atmospheric circulation

Revised ☐

The distribution of biomes can be explained by the tricellular model of atmospheric circulation (Figure 2.30), which shows how differences in pressure and corresponding wind lead to differences in precipitation at different **latitudes**. The model explains why rainfall is high at the equator and at 60° north and south.

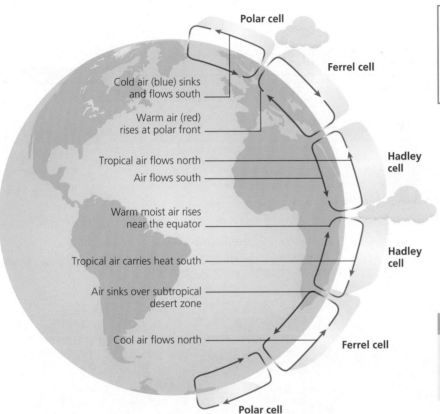

> **Keyword definition**
>
> **Latitude** – The angular distance from the equator (north or south of it) as measured from the centre of the Earth (usually in degrees).

Figure 2.30 The tricellular model of atmospheric circulation

The Hadley cell controls weather over the tropics, where the air is warm and unstable. High levels of insolation at the equator heat up the air. Hot air rises, creating the Hadley cell.

As hot air rises, it cools and condenses, forming large clouds that lead to the heavy rainfall that is characteristic of tropical rainforest. The pressure at the equator is low because air is rising.

The air from the Hadley cell cools as it travels away from the equator, where it meets air from the Ferrel cell. The cooled air descends at 30° north and south of the equator.

Pressure here is high. The air is dry, and so it is in these locations that the desert biome is found.

Air either returns to the equator at ground level or travels towards the poles as warm winds. Where the warm air travelling north and south hits the colder polar winds, at approximately 60° north and south of the equator, it rises because it is less dense. This creates an area of low pressure.

As the air rises, it cools and condenses, forming clouds. Precipitation results, so this is where temperate forest biomes are found.

> **Expert tip**
>
> The tricellular model is made up of the polar cell, the Ferrel cell in mid-latitudes and the Hadley cell in the tropics.

> **Expert tip**
>
> Downward air movement creates high pressure. Upward air movement creates low pressure and cooling air that leads to increased cloud formation and precipitation.

> **Expert tip**
>
> The tricellular model of atmospheric circulation explains the distribution of precipitation and temperature and how they influence structure and relative productivity of different terrestrial biomes.

8 Explain how different limiting factors will determine productivity in *two* contrasting biomes. [4]

9 Explain why productivity in tundra is low. [3]

10 Tropical rainforest has a mean NPP of 2.20 kg m^{-2} yr^{-1}, and tundra 0.14 kg m^{-2} yr^{-1}. When NPP per kg biomass per year is calculated, tropical rainforest has a value of 0.049, and tundra a value of 0.233. Compare and explain these data. [4]

The effect of climate change on biomes

Revised

The distribution of biomes is dependent on climate (page 47). Changes in patterns of rainfall associated with climate change, and alterations in land and sea surface temperature, will lead to shifts in the distribution of biomes (Figure 2.31; see also Chapter 7, pages 137–8).

Expert tip

You need to be able to discuss the impact of climate change on biomes.

The colour of each semicircle indicates the retracting biome (top for North America, Europe, Asia; bottom for Africa and New Zealand) and the expanding biome (bottom for North America, Europe, Asia; top for Africa and New Zealand

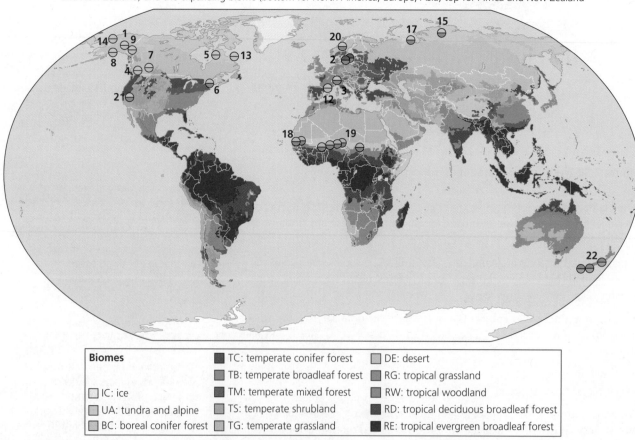

Biomes	■ TC: temperate conifer forest	□ DE: desert
	■ TB: temperate broadleaf forest	■ RG: tropical grassland
□ IC: ice	■ TM: temperate mixed forest	■ RW: tropical woodland
■ UA: tundra and alpine	■ TS: temperate shrubland	■ RD: tropical deciduous broadleaf forest
■ BC: boreal conifer forest	□ TG: temperate grassland	■ RE: tropical evergreen broadleaf forest

Figure 2.31 Observed biome shifts during the twentieth century

Succession

Revised

Changes in the community of organisms frequently cause changes in the physical environment that allow another community to become established and replace the former through competition. Often, but not inevitably, the later communities in such a sequence or **sere** are more complex than those that appear earlier.

The formation of an ecosystem from, for example, bare rock is called **primary succession**:

- **Pioneer species** arrive (e.g. lichens, mosses, bacteria).
- As pioneers die, soil is created.
- New species of plant arrive that need soil to survive – these displace pioneer species.
- Growth of plants causes changes in the environment (e.g. light, wind, moisture).
- Growth of roots enables soil to be retained; nutrients and water in the soil increase.
- Nitrogen-fixing plants arrive, adding nitrates to the soil.
- Soil depth increases further, allowing shrubs and other taller plants to arrive.
- Animal species arrive as species of plant they rely on become established.
- A **climax community** is established (Figure 2.32).

Succession in areas that already have soil is called **secondary succession**.

Diversity changes through succession (see pages 53–54):

- Greater habitat diversity leads to greater species and genetic diversity.
- Complex (i.e. climax) ecosystems have a wider variety of nutrient and energy pathways, which provide stability Chapter 1 – page 14, and pages 54–55.

> **Keyword definition**
>
> **Succession** – The orderly process of change *over time* in a community.

> **Expert tip**
>
> During succession, the patterns of energy flow, gross and net productivity, diversity, and mineral cycling change over time. Through the succession, greater habitat diversity leads to greater species and genetic diversity.

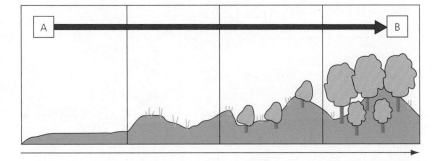

Figure 2.32 Succession in a sand dune ecosystem. The *x*-axis represents either time or distance from the sea. A is a pioneer community and B a climax community

Table 2.7 Comparison of pioneer and climax communities

Pioneer communities	Climax communities
Contain the first organisms to colonise a new environment	The end-point of ecological succession
Dominated by *r*-strategists (see below)	Dominated by *K*-strategists (see below)
Simple in structure, with low diversity	Complex in structure, with high diversity
Tolerate harsh conditions, for example strong light/low nutrient levels	Characteristics determined by climate and soil
Example: community of lichens covering bare rock	Example: mature rainforest ecosystem

> **Expert tip**
>
> There is no one climax community, but rather a set of alternative stable states for a given ecosystem. These depend on the climatic factors, the properties of the local soil and a range of random events that can occur over time.

Keyword definitions

Sere – The set of communities that succeeds another over the course of succession at a given location.

Pioneer community – The first stage of an ecological succession that contains hardy species able to withstand difficult conditions.

Climax community – A community of organisms that is more or less stable, and that is in equilibrium with natural environmental conditions such as climate. It is the end point of ecological succession.

Expert tip

You need to be able to describe the process of succession in a given example and explain the general patterns of change in communities undergoing succession. Named examples of organisms from the pioneer, intermediate and climax communities should be provided.

Examples of succession, from pioneer community to climax community, are shown in Figures 2.33 and 2.34.

aquatic plants (e.g. water lilies) → reeds → low woodland species (e.g. willow)

Figure 2.33 Succession in freshwater

Expert tip

If you are asked to give an example of a succession, include organisms from the pioneer community, the climax community and the seral stages in between.

mosses and lichens → grasses and herbs → shrubs (e.g. birch) → woodland

Figure 2.34 Succession in an abandoned quarry

■ **QUICK CHECK QUESTIONS**

23 Define the term *succession*.

24 Distinguish between the terms *succession* and *zonation*.

25 Define the term *climax community*.

26 Outline the differences between pioneer and climax communities.

27 Give an example of a succession. How would this succession change over time?

Density-dependent and density-independent factors

Revised

- **Density-dependent factors** are limiting factors that are related to population density. They are biotic factors (e.g. competition for resources) that limit population growth.
- Internal density-dependent factors might include fertility or size of breeding territory.
- External density-dependent factors might include predation or disease.
- **Density-independent factors** are abiotic and do not depend on the size of the population.
- Density-dependent factors operate as negative feedback mechanisms, leading to stability or regulation of the population.

When the population growth of a species is determined by limiting factors it reaches a carrying capacity, *K*. Populations of these *K*-selected species, or *K*-strategists, are controlled by density-dependent factors. Examples include elephants, humans and whales.

Other species have high reproductive (*r*) rates and are known as *r*-selected species or *r*-strategists. They are controlled by density-independent factors. Examples include house flies, cockroaches and mosquitoes.

Keyword definitions

K-strategists – Species that usually concentrate their reproductive investment in a small number of offspring, thus increasing their survival rate and adapting them for living in long-term climax communities.

r-strategists – Species that tend to spread their reproductive investment among a large number of offspring so that they are well adapted to colonise new habitats rapidly and make opportunistic use of short-lived resources.

Expert tip

You need to be able to distinguish the roles of *r*- and *K*-selected species in succession.

Table 2.8 Comparison of *r*- and *K*-strategist species

r-strategist	*K*-strategist
Smaller in size	Larger in size
Early maturity and reproduction	Late maturity and delayed reproduction
Little or no parental care	Large amount of parental care
Large number of offspring	Few offspring
Little investment in individual offspring	High investment in individual offspring
Rapid growth and development	Slow development
Shorter life	Longer life
Thrive in unstable environments	Thrive in stable environments
J-population growth curve	S-population growth curve
Generalist species	Specialist species

Changes through a succession

Revised

Table 2.9 shows the changes that occur though a succession, from point A through to B in Figure 2.32 (page 51).

Table 2.9 Features of an ecosystem at early and late stages during the process of succession

	Pioneer community	Climax community
Amount of organic matter	Small	Large
Soil depth	Shallow	Deep
Soil quality	Immature/little organic material	Mature/much organic matter
Nutrients	External	Internal
Nutrient cycles	Open system	Closed system
Nutrient conservation	Poor	Good
Role of detritus	Small	Large
Stability	Poor	Good
Niches	Wide	Narrow
Species richness	Low	High
Diversity	Low	High
Size of organisms	Small	Large
Life cycles	Simple	Complex
Growth form	*r*-selected species	*K*-selected species

Productivity also changes through a succession:

■ In pioneer communities, gross productivity is low due to the initial conditions and the low density of producers. The proportion of energy lost through community respiration is also relatively low. Net productivity is therefore high: the system grows and biomass accumulates.

■ In later stages there is an increased consumer community.

Common mistakes

It is not enough to say that 'productivity increases through a succession'. GPP increases but NPP decreases as community respiration increases.

■ In climax communities gross productivity may be high, but this is balanced by respiration. The **production:respiration (P:R) ratio** is a measure of productivity relative to respiration. Net productivity approaches 0 in climax communities and the P:R ratio approaches 1.

Climax communities

Climatic and edaphic (soil) factors determine the nature of a climax community. Climax communities are more stable than earlier seral stages because:

■ they contain more complex food webs – this provides more stability because if one organism goes extinct it can be replaced by another
■ negative feedback mechanisms lead to steady-state equilibrium
■ each seral stage helps to create a deeper and more nutrient-rich soil, so allowing larger plants to grow
■ a climax community is more productive, providing more energy to support consumers and decomposers
■ increased biomass leads to an increased number of niches, which increases species and genetic diversity, resulting in greater stability.

Ecosystem stability and succession

Human factors frequently modify succession through, for example, burning, agriculture, deforestation, hunting, grazing and habitat clearance. They often simplify ecosystems, rendering them unstable – for example, North American wheat farming, a modified system, compared with tall grass prairie, a climax community.

An ecosystem's capacity to survive change may depend on **diversity** and **resilience** (Chapter 1, pages 14–15).

■ **Diversity** refers to the number and relative abundance of species present (see pages 60–61).
■ **Resilience** refers to the tendency of a system to avoid tipping points, and maintain stability through steady-state equilibrium.

Large storages and high diversity mean that a system is less likely to reach a tipping point and move to a new equilibrium. A resilient ecosystem is able to maintain its structure, ecological functions and processes. Humans can affect the resilience of systems through reducing storages and diversity (Chapter 1, page 15). Tall grass prairie has higher diversity and greater resilience, for example, than a monoculture created by wheat farming:

■ The greater diversity of tall grass prairie creates a more complex food web that is more resistant to change, and therefore a more stable ecosystem.
■ In complex, diverse ecosystems, certain species perform essential functions. Bird predation, for example, can maintain low numbers of insects, reducing levels of insect herbivory and therefore increasing productivity. Pollinators are also important for the functioning of ecosystems and without them many plant species would not be able to reproduce.
■ There is a greater store of seeds in the soil of tall grass prairie so that recovery can happen more quickly following disturbance.
■ Tall grass ecosystems have deep soils with large stores of nutrients that are effectively cycled, compared with monocultures where soils may be poor and need to be maintained by regularly applying fertilisers.

Disturbance can stop the process of succession so that the climax community is not reached. Interrupted succession is known as a **plagioclimax**. Human activity can have various effects on climax communities (Figure 2.35):

■ Decrease in productivity through the removal of primary producers.
■ Reduction in producers, leading to reduced habitat diversity and fewer niches, which threatens more specialised species.

> **Expert tip**
>
> You need to be able to discuss the link between ecosystem stability, succession, diversity and human activity.

Figure 2.35 Interrupting a succession through human activity

■ Deterioration in abiotic factors, leading to harsher conditions that fewer species can adapt to.
■ Species extinction, leading to shorter food webs.
■ A less complex community, leading to decreased stability.

Expert tip

Various factors can divert the progression of succession to an alternative stable state by modifying the ecosystem, such as the use of fire, agriculture, grazing pressure or resource use (such as deforestation). This diversion may be more or less permanent depending upon the resilience of the ecosystem.

Expert tip

You need to be able to discuss the factors that could lead to alternative stable states in an ecosystem.

■ QUICK CHECK QUESTIONS

28 Describe the process of succession.

29 Distinguish between *primary* and *secondary succession*.

30 An ecosystem's capacity to survive change may depend on diversity and resilience. Define the terms *diversity* and *resilience*.

31 Explain how diversity and resilience will affect the ability of a named ecosystem to respond to disturbance.

Zonation

Revised

The main biomes display **zonation** in relation to latitude and climate. Plant communities can also display zonation with altitude on a mountain (Figure 2.36), or around the edge of a pond in relation to soil moisture.

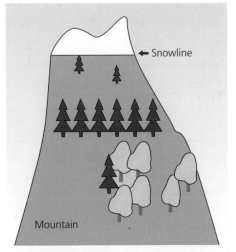

← Snowline

Mountain

Figure 2.36 Zonation of vegetation on a mountain

Keyword definition

Zonation – The arrangement or patterning of plant communities or ecosystems into parallel or sub-parallel bands in response to change, over a distance, in some environmental factor.

CASE STUDY

MOUNT KINABALU, BORNEO

Mount Kinabalu in Malaysian north Borneo (Figure 2.37) is an example of altitudinal zonation where plant communities vary from tropical rainforest at low levels to alpine communities near the summit. Zonation is caused by changes in temperature, from 26°C in the rainforest zone to 0°C at the summit.

Figure 2.37 Mount Kinabalu: an example of altitudinal zonation

EXAM PRACTICE

11 Explain the relationship between succession and stability. [6]

12 a Define the term *succession*. [1]

   b Explain how diversity changes through a succession. [5]

   c Using the terms *diversity* and *resilience*, outline the effects of disturbance on a named ecosystem. [6]

13 Compare the strategies that pioneer species and climax community species are likely to have in terms of specific growth rate, parental care and competitive advantage. [4]

14 Explain what is meant by the terms *ecological succession*, *pioneer community* and *climax community*. [6]

15 Describe and explain how gross primary productivity changes during the stages of succession. [6]

16 Distinguish between the terms *succession* and *zonation*. [6]

Expert tip

The concept of succession, occurring *over time*, should be carefully distinguished from the concept of zonation, which refers to a *spatial pattern*. An exam question may ask you to distinguish between the two.

Expert tip

You need to be able to interpret models or graphs related to succession and zonation.

2.5 Investigating ecosystems

Revised ☐

SIGNIFICANT IDEAS

- The description and investigation of ecosystems allows for comparisons to be made between different ecosystems and for them to be monitored, modelled and evaluated over time, measuring both natural change and human impacts.
- Ecosystems can be better understood through the investigation and quantification of their components.

Expert tip

You need to be able to design and carry out ecological investigations.

Expert tip

The study of an ecosystem requires that it be named and located; for example, Danum Valley Field Centre in Sabah, Malaysia – a tropical lowland rainforest.

Measuring abiotic components of the system

Revised ☐

Standardised methods for measuring abiotic factors (see page 29) are needed to compare ecosystems (Table 2.10).

Table 2.10 The measurement of abiotic factors in ecosystems

Abiotic factor	How is it measured?	Evaluation
Wind speed*	Anemometer (Figure 2.38)	Gusty conditions can lead to large variations in data
Temperature+	Thermometer	Problems in data if temperature not taken from consistent depth
Light+	Lightmeter	Cloud cover changes light intensity, as does shading from plants or lightmeter operator
Flow velocityx	Flowmeter (Figure 2.39)	Readings must be taken from same depth; water flow can vary due to rainfall/ice melt
Wave action†	Dynamometer	Changes in wave strength during a day and over a monthly period affect results
Turbidity†	Secchi disc	Reflections off water reduce visibility; measurements are subjective
Dissolved oxygenx	Oxygenmeter	Possible contamination from oxygen in air when using oxygenmeter
Soil moisture*	Evaporate water; soil moisture probes	If soil is too hot when evaporating water, organic content can also burn off

Type of ecosystem where technique is mainly used: *terrestrial; xfreshwater; †marine; +all three

Expert tip

You need to be able to evaluate methods to measure at least three abiotic factors in an ecosystem.

Expert tip

You need to be able to evaluate sampling strategies.

Figure 2.38 Using an anemometer to measure wind speed in a shingle ridge succession.

Figure 2.39 Using a flowmeter to measure water speed in a forest river

Abiotic factors can vary from day to day and season to season. Electronic data-loggers overcome many of the limitations shown by abiotic measuring devices:

- They provide continuous data over a long period of time.
- They make data more representative of the area being sampled.
- More data can be collected, making results more reliable.

> **Common mistakes**
>
> 'Climate' and 'temperature' are sometimes used interchangeably. These terms are not the same: climate includes rainfall, humidity and wind speed as well as temperature.

> **Expert tip**
>
> If you are using sampling methods as examples in exam questions, use specific examples and avoid vague answers. If you have been on a field trip, knowledge of techniques gained through fieldwork should be used.

■ QUICK CHECK QUESTIONS

32 Identify an abiotic factor found in a freshwater ecosystem. Outline how you would measure this factor.

33 Identify an abiotic factor found in a marine ecosystem. Outline how this factor would vary with depth.

34 Identify an abiotic factor found in a terrestrial ecosystem. Evaluate the technique used to measure this factor.

Identifying organisms in ecosystems

Revised ☐

In ecological projects, it is important to correctly identify organisms being studied. Organisms in an ecosystem can be identified:

- by comparing specimens with those in a herbarium or with museum specimen collections
- using DNA profiling techniques to identify differences between specimens (a more accurate way of determining the identity of an organism than just using its physical appearance)
- using dichotomous keys (see below).

A **dichotomous key** is organised in steps, with two options given at each step. The two options identify contrasting features of the species.

Worked example

Construct a key for the following animals: spider, beetle, monkey, gibbon, rhinoceros, eagle, snake, frog, leopard, butterfly, kangaroo and dolphin.

1	a	Animal is a vertebrate	*go to 4*
	b	Animal is an invertebrate	*go to 2*
2	a	Animal has fewer than 8 legs	*go to 3*
	b	Animal has 8 legs	**spider**
3	a	Animal has two hardened wing cases	**beetle**
	b	Animal does not have modified wings	**butterfly**
4	a	Animal is warm blooded (endothermic)	*go to 6*
	b	Animal is cold blooded (ectothermic)	*go to 5*
5	a	Animal has legs	**frog**
	b	Animal does not have legs	**snake**
6	a	Animal is adapted for life in water	**dolphin**
	b	Animal is not adapted for life in water	*go to 7*
7	a	Animal has wings	**eagle**
	b	Animal does not have wings	*go to 8*
8	a	Animal has a placenta	*go to 9*
	b	Animal does not have a placenta	**kangaroo**
9	a	Animal has fur	*go to 10*
	b	Animal does not have fur	**rhinoceros**
10	a	Animal has claws on feet	**leopard**
	b	Animal does not have claws on feet	*go to 11*
11	a	Animal has a tail	**monkey**
	b	Animal does not have a tail	**gibbon**

Limitations of keys include:

- The organism might not be in the key.
- Terminology can be difficult.
- There might not be a key available for the organisms under investigation.
- Some features cannot be easily established in the field – for example, whether an animal has a placenta or not, or whether an animal is endothermic or ectothermic.

Measuring biotic components of the system

Revised

Standardised methods are needed to compare biotic components of ecosystems with one another. Such studies also allow ecosystems to be monitored and evaluated over time, and for the effects of human disturbance to be understood.

Expert tip
When carrying out fieldwork you must follow the IB ethical practice guidelines and IB animal experimentation policy: that is, animals and the environment should not be harmed during your work.

Methods for estimating the abundance of organisms

Revised

The way in which the abundance of an organism is measured depends on whether it is **motile** or **non-motile**.

■ Lincoln index

This technique is known as the capture–mark–release–recapture method. It is used for estimating the population size of motile animals.

Expert tip
Keys can also be shown as a diagram, with branches representing each step.

Expert tip
You need to be able to construct simple identification keys for up to eight species.

Expert tip
You need to be able to evaluate methods for measuring or estimating populations of motile and non-motile organisms.

- Organisms are captured, marked, released and then recaptured.
- Marking varies according to the type of organism. For example, wing cases of insects can be marked with pen, snails with paint, and fur clippings used for mammals.
- Markings must be difficult to see – high visibility increases predation risk.
- The number of individuals of a species is recorded at each stage.
- The total population size is estimated using the following equation:

$$N = \frac{n_1 \times n_2}{m}$$

Where: N = total population of animals in the study site; n_1 = number of animals captured (marked and released) on first day; n_2 = number of animals recaptured on second day; m = number of marked animals recaptured on second day.

■ Quadrat methods

Quadrats are used for estimating the abundance of plants and non-motile animals.

- **Percentage frequency** is the percentage of quadrats in an area in which at least one individual of the species is found. It is calculated by taking the number of occurrences and dividing by the number of possible occurrences; for example, if a plant occurs in 3 out of 100 squares in a grid quadrat, then the percentage frequency is 3%.
- **Percentage cover** is the proportion of a quadrat covered by a species, measured as a percentage (Figure 2.40). It is worked out for each species present. Estimates can be made by dividing the quadrat into a 10 × 10 grid (100 squares), where each square is 1% of the total area covered.
- **Population density** is the number of individuals of each species per unit area. It is calculated by dividing the number of organisms by the total area of the quadrats.

Figure 2.40 A quadrat being used to estimate percentage cover of seaweed on a rocky shore

The sampling system used depends on the areas being sampled:

- **Random sampling** is used if the same habitat is found throughout the area.
- **Stratified random sampling** is used in two areas different in habitat quality.
- **Systematic sampling** is used along a **transect** where there is an environmental gradient.

Method for estimating the biomass of trophic levels

Biomass is calculated to indicate the total energy within a trophic level.

- Biomass is a measure of the organic content of organisms.
- Water is not an organic molecule, and its amount varies from organism to organism, so water is removed before biomass is measured. This is called **dry weight biomass**.

Revised ☐

- One criticism of the method is that it involves the killing of living organisms (although not all the organisms in an area need to be sampled – see below).
- Problems exist with measuring biomass of very large plants such as trees, and with roots and underground biomass.

Calculating dry weight biomass

To obtain quantitative samples, biological material is dried to constant weight:

- The sample is weighed in a container of known weight.
- The sample is put in a hot oven (80°C).
- After a specific length of time the sample is reweighed.
- The sample is put back in the oven.
- This is repeated until the same mass is recorded from two successive readings.
- No further loss in mass indicates that water is no longer present.

Biomass is recorded per unit area (e.g. per metre squared) so that trophic levels can be compared. Not all organisms in an area need to be sampled:

- The mass of one organism, or the average mass of several organisms, is taken.
- This mass is multiplied by the total number of organisms to estimate total biomass.
- This is called an **extrapolation technique**.

> **Expert tip**
>
> Data from methods for estimating biomass can be used to construct ecological pyramids.

> **Expert tip**
>
> You need to be able to evaluate methods for estimating biomass at different trophic levels in an ecosystem.

> **Expert tip**
>
> To estimate the biomass of a primary producer, all the vegetation, including roots, stems and leaves, is collected within a series of 1 m × 1 m quadrats. The dry weight method is carried out and average biomass calculated.

■ QUICK CHECK QUESTIONS

35 Describe and evaluate methods for measuring *three* abiotic factors in a forest ecosystem.

36 Explain the difference between percentage frequency and percentage cover.

37 Which data are needed to estimate the size of an animal population? Write the equation needed to calculate population size.

38 Explain how biomass is calculated.

Diversity and the Simpson's diversity index

Revised

Species diversity refers to the number of species and their relative abundance (see Topic 4). It can be calculated using **diversity indices**.

Species diversity can be calculated using the **Simpson's diversity index**, using the equation:

$$D = \frac{N(N-1)}{\sum n(n-1)}$$

Where: D = Simpson's index; N = total number of organisms of all species found; n = number of individuals of a particular species.

Index values are relative to each other and not absolute, unlike measures of, say, temperature, which are on a fixed scale.

A **diversity index** is a numerical measure of species diversity calculated by using both the number of species (species richness) and their relative abundance.

- Comparisons can be made between areas containing the same type of organism in the same ecosystem.
- A high value of D suggests a stable and ancient site, where all species have similar abundance (or 'evenness').
- A low value of D could suggest disturbance through, say, logging, pollution, recent colonisation or agricultural management, where one species may dominate.

> **Keyword definition**
>
> **Diversity** – A generic term for heterogeneity (i.e. variation or variety). The scientific meaning of diversity becomes clear from the context in which it is used; it can refer to heterogeneity of species or habitat, or to genetic heterogeneity.

> **Expert tip**
>
> Similar habitats can be compared using D; a lower value in one habitat may indicate human impact. Low values of D in the Arctic tundra, however, may represent stable and ancient sites.

> **Expert tip**
>
> You need to be able to calculate and interpret data for species richness and diversity.

■ QUICK CHECK QUESTION

39 One habitat has a Simpson's index of 1.83 and another has an index of 3.65. What do these values tell you about each habitat?

Worked example

The table below contains data from two different habitats. Total number of species (= 'species richness') and total number of individuals is the same in each case. Calculate the diversity of both habitats and comment on the differences between the habitats.

Species found	Number found in habitat X	Number found in habitat Y
A	10	3
B	10	5
C	10	2
D	10	36
E	10	4
Number of species =	**5**	**5**
Number of individuals =	**50**	**50**

Simpson's index must be calculated for each habitat. This can be done using a table to calculate components of the index:

Species	Numbers (n) found in habitat X	$n(n-1)$	Numbers (n) found in habitat Y	$n(n-1)$
A	10	10(9) = **90**	3	3(2) = **6**
B	10	10(9) = **90**	5	5(4) = **20**
C	10	10(9) = **90**	2	2(1) = **2**
D	10	10(9) = **90**	36	36(35) = **1260**
E	10	10(9) = **90**	4	4(3) = **12**
	$\Sigma n(n-1)$	**450**	$\Sigma n(n-1)$	**1300**

Species diversity for each habitat:

- Habitat X:

$$D = \frac{50(49)}{450} = \frac{2450}{450} = \textbf{5.44}$$

- Habitat Y:

$$D = \frac{50(49)}{1300} = \frac{2450}{1300} = \textbf{1.88}$$

What do these values say about each habitat?

- Greater 'evenness' between species in habitat X.
- Less competition due to non-overlapping niches in habitat X.
- One species does not dominate in X, reflecting greater habitat complexity/ more niches.
- Habitat Y – less complex with fewer/overlapping niches, where one species can dominate, leading to lower diversity.

Expert tip

You need to be able to draw graphs to illustrate differences in species diversity in a community over time, or between communities.

Measuring changes along an environmental gradient

Revised

Ecological gradients are found where two ecosystems meet or where an ecosystem ends. Abiotic and biotic factors change along an ecological gradient (see page 56–59 for a review of sampling techniques).

Transects are used to measure changes along the gradient; this ensures that all parts of the gradient are measured (Figure 2.41):

■ The whole transect can be sampled (a **continuous** transect) *or* samples can be taken at points of equal distance along the transect (an **interrupted** transect).
■ A **line transect** is the simplest transect, where a tape measure is laid out in the direction of the gradient. All organisms touching the tape are recorded.
■ A **belt transect** allows more samples to be taken – a band usually between 0.5 m and 1 m is sampled along the gradient.

Quadrats can be used to sample at regular intervals along a transect (see page 59 for further information):

■ **Frame quadrats** are empty frames of known area (e.g. 1 m^2).
■ **Grid quadrats** are frames divided into 100 small squares.
■ **Point quadrats** are made from a frame with 10 holes, inserted into the ground by a leg (Figure 2.42). They are used for sampling vegetation that grows in layers. A pin is dropped through each hole in turn and the species touched are recorded. The total number of pins touching each species is converted to percentage frequency data (i.e. if a species touches 7 out of the 10 pins it has 70% frequency).

Figure 2.41 Sampling along an environmental gradient on a shingle ridge succession. A tape measure is laid out at 90° to the sea and abiotic and biotic factors measured at regular intervals along it

Zonation can be measured by recording biotic and abiotic factors at fixed heights along a transect:

■ A cross staff is used to move a set distance (e.g. 0.6 m) vertically up the transect (Figure 2.43).
■ The staff is set vertically and a point measured horizontally from an eye-sight 0.6 m from the base of the staff.
■ Biotic and abiotic factors are measured at each height interval.

Figure 2.42 Using a point quadrat

Figure 2.43 A cross staff being used to relocate quadrats at regular height intervals along a rocky shore. This allows zonation on the shore to be studied

■ **QUICK CHECK QUESTIONS**

40 What is meant by the term *ecological gradient*?

41 Outline how you would measure changes in abiotic factors along an environmental gradient.

42 Describe *three* different methods for recording biotic factors along a belt transect.

43 Describe how you would collect data to show zonation in an ecosystem.

Expert tip

You need to be able to evaluate methods to investigate the change along an environmental gradient and the effect of a human impact on an ecosystem.

Expert tip

Studying both biotic and abiotic factors allows research questions such as: how do abiotic factors affect the distribution of organisms in ecosystems? Different species can be expected to be found at different locations along the gradient as they will be adapted to different conditions.

Common mistake

Do not confuse the terms biotic and abiotic – biotic refers to the living parts of the ecosystem and abiotic to the non-living parts.

Measuring changes due to a specific human activity

Revised

Human activities can change abiotic and biotic components of an ecosystem. Human impacts include toxins from mining activity, landfills, eutrophication, effluent, oil spills and overexploitation.

Techniques outlined in this section (pages 56–63) can be used to investigate ecosystems and to see how human activities are affecting them.

CASE STUDY

DEFORESTATION IN THE AMAZON

Scientists are measuring the effects of human activity on the Amazon rainforest.

■ Logging is causing large areas of forest to be lost.

■ Deforestation is being carried out to provide timber, and to clear land for agriculture and housing.

■ Remaining areas can become fragmented, forming islands of habitat.

Satellite photos (Figure 2.44) can be used to monitor the amount of deforestation taking place:

■ This method is very reliable, can cover a large area, and monitors change over time.

■ The visual impact of the photos is useful for motivating action against logging.

Abiotic and biotic factors can be measured:

Figure 2.44 Satellite photo showing deforestation in the Amazon rainforest

■ These must be taken in both undisturbed and disturbed habitats so that comparisons can be made.

■ Habitat islands of different sizes will have different environmental conditions, and so a variety of different-sized patches must be measured.

■ Samples must be repeated so that data are reliable.

■ Abiotic factors can include temperature, humidity and sunlight.

■ Biotic factors can include the species of plants and animals present, and the population sizes of selected indicator species.

■ Where environmental gradients are present, factors must be measured along the full extent of the gradient so that valid comparisons can be made.

■ Factors must be measured over a long period of time to take into account daily and seasonal variations.

See pages 56–59 for description and evaluation of methods for measuring abiotic and biotic factors.

EXAM PRACTICE

17 a Name an ecosystem you have studied and state *one* abiotic factor you can measure. [1]

   b Outline how you would measure changes in the abiotic factor over time. [2]

   c Explain why differences in the abiotic factor named in part **a** might affect diversity between two sites. [2]

18 With the help of examples, suggest a research question that would connect the abiotic and biotic components of an ecosystem. [2]

19 Outline and evaluate a method for estimating the abundance of a named plant species in a named ecosystem. [3]

20 Describe and evaluate a method for estimating the abundance of rhinos in an African national park. [4]

21 a Explain why abundance of organisms might be of importance in estimating the diversity of an ecosystem. [2]

   b Explain how you would compare the diversity of *two* different ecosystems. [5]

22 a Calculate the Simpson's diversity index for habitat B using data from the table below. Use the formula:

$$D = \frac{N(N-1)}{\Sigma n(n-1)}$$ [2]

Species found	Number of individuals found in each habitat				
	Habitat A	Habitat B	Habitat C	Habitat D	Habitat E
Species 1	25	50	80	97	100
Species 2	25	30	10	1	0
Species 3	25	15	5	1	0
Species 4	25	5	5	1	0
Simpson's index (*D*)	4.13		1.37	1.04	1.00

   b i) Describe and explain the differences between the habitats shown in the table.

      ii) Suggest reasons for the different values of Simpson's index recorded in different habitats. [4]

23 a Describe how changes in species composition along an environmental gradient might affect a named abiotic factor. [2]

   b Outline and evaluate methods that you could use in the field to gather evidence for i) species composition and ii) your suggestion in part **a**. [6]

24 Describe a method for measuring changes in abiotic components in a named ecosystem affected by human activity. [5]

25 a Suggest *two* characteristics of species that usually make them suitable for sampling with quadrats. [2]

   b Explain how you might ensure that the quadrats were placed at random. [2]

Expert tip

Measurements should be repeated to increase reliability of data. The number of repetitions required depends on the factor being measured.

■ QUICK CHECK QUESTIONS

44 Outline how human activities can change the abiotic and biotic components of an ecosystem.

45 Describe and evaluate methods for measuring changes in abiotic and biotic components of an ecosystem due to a named specific human activity.

Topic **3** Biodiversity and conservation

3.1 An introduction to biodiversity

> **SIGNIFICANT IDEAS**
> - Biodiversity can be identified in a variety of forms, including species diversity, habitat diversity and genetic diversity.
> - The ability to both understand and quantify biodiversity is important to conservation efforts.

Biodiversity

Biodiversity is a broad concept encompassing the total diversity of living systems, which includes the **diversity of species**, **habitat diversity** and **genetic diversity**.

Conservation efforts rely on the quantification of biodiversity to provide an understanding of natural systems and the effect of human activities on them.

■ Species, habitat and genetic diversity

In each cell, **DNA** contains the genetic information that codes for the structure and characteristics of an organism. The DNA is divided into sections that code for particular characteristics in a species – these sections are called **genes**.

The term **gene pool** refers to all the different genes in a population. **Species** with a small gene pool (i.e. low genetic diversity) are more at threat from extinction than those with a larger gene pool. In a large gene pool, it is more likely that alleles exist that will help the species survive any change in their environment. A smaller gene pool means that adaptive genes are less likely to exist and a species is less able to adapt and survive (Figure 3.1).

Figure 3.1 Cheetahs have a small gene pool – such species are prone to extinction

> ■ **QUICK CHECK QUESTIONS**
> 1 Define the term *biodiversity*.
> 2 Explain the difference between *species richness* and *species diversity*.
> 3 Outline the differences between *habitat diversity* and *genetic diversity*.

Keyword definitions

Biodiversity – The amount of biological or living diversity per unit area. It includes the concepts of species diversity, habitat diversity and genetic diversity.

Species – A group of organisms that interbreed and are capable of producing fertile offspring.

Species diversity – The variety of species per unit area. This includes both the number of species present and their relative abundance.

Habitat diversity – The range of different habitats in an ecosystem or biome. Conservation of habitat diversity usually leads to the conservation of species and genetic diversity.

Genetic diversity – The range of genetic material present in a population of a species.

Expert tip

You need to be able to distinguish between biodiversity, diversity of species, habitat diversity and genetic diversity.

Diversity indices

Revised ☐

Species diversity in communities is a product of two variables: the number of species (richness) and their relative proportions (evenness). Species richness in an area can be high, but diversity low (Figure 3.2). Species diversity is a better measure of ecosystem quality and health than species richness (see Case study, below).

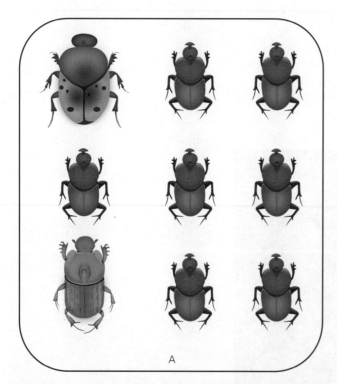   

Figure 3.2 Species richness vs species diversity. The species richness is the same in both ecosystems (ecosystem A and ecosystem B – three species of beetle in each). In ecosystem B, diversity is higher than ecosystem A because evenness is higher. One species dominates in ecosystem A, indicating a less complex ecosystem

Diversity indices, such as Simpson's index (Chapter 2, pages 60–61), can be used to describe and compare communities. Diversity indices can be used to assess whether the impact of human development on ecosystems is sustainable or not. When comparing communities that are similar:

■ low diversity could indicate pollution, eutrophication or recent colonisation of a site
■ the number of species present in an area is often indicative of general patterns of biodiversity, although disturbed sites often have artificially increased species richness due to the mixing of habitats that are usually spatially separate in undisturbed sites.

CASE STUDY

BEETLE COMMUNITIES IN BORNEO

Pitfall traps were used to sample dung beetle communities in the tropical rainforest of Borneo. Pitfalls were made from pots buried in the soil, baited with primate dung to attract beetles (Figure 3.3). Pitfalls are a method of sampling mobile populations (see Chapter 2, page 58). Pitfalls were used to collect beetles in primary (undisturbed) forest by a large river (riverine forest), primary forest further in the interior, logged forest (areas harvested for timber) and plantation (one species of tree grown to produce a local crop).

Results from the investigation are shown in Table 3.1.

Table 3.1 Results from pitfall trapping of dung beetles in the rainforest of Sabah, Malaysian Borneo

Location	Species richness	Diversity	Dominance index
Riverine primary forest	48	1.92	0.54
Primary interior forest	42	2.63	0.24
Logged forest	46	1.59	0.66
Plantation	29	1.78	0.52

Figure 3.3 Baited pitfall traps used to collect dung beetles

Species richness is the total number of species collected in each forest type. Diversity was calculated for each location, and also a dominance index. The dominance index is a measure of the degree to which a few species dominate the community – high dominance indicates low evenness.

Results show that riverine forest had the highest species richness – this is because it is an area where several habitats meet, increasing the number of species sampled. This forest had a lower diversity than interior primary forest, due to the dominance of one species and the corresponding lower evenness.

Logged forest had high species richness but lower diversity compared with undisturbed forest. Logged forest contains a mixture of species usually separated along environmental gradients in primary forest: riverine species and those found in the canopy move into logged areas, increasing species richness artificially. The lower species diversity in logged forest indicates a simplified ecosystem where some species dominate (indicated by a high dominance measure). Plantation forest has the lowest species richness and also a low diversity, indicating a loss of primary forest species and a much simpler ecosystem compared with primary rainforest.

Species richness is a poor indicator of habitat disturbance when used on its own. Diversity indices, and measures of evenness, more clearly show changes to community structure as a result of human impact.

Species found	Number of individuals found in each habitat				
	Habitat A	**Habitat B**	**Habitat C**	**Habitat D**	**Habitat E**
Species 1	25	50	80	97	100
Species 2	25	30	10	1	0
Species 3	25	15	5	1	0
Species 4	25	5	5	1	0
Simpson's diversity index (*D*)	4.13	2.79	1.37	1.04	1.00

1a Explain the differences between species richness and diversity. [2]

b Explain how diversity indices can be used to assess the impact of human activities on ecosystems. [3]

3.2 Origins of biodiversity

Revised ☐

SIGNIFICANT IDEAS

- Evolution is a gradual change in the genetic character of populations over many generations, achieved largely through the mechanism of natural selection.
- Environmental change gives new challenges to specie, which drive the evolution of diversity.
- There have been major mass extinction events in the geological past.

Evolution

Revised ☐

Biodiversity arises from evolutionary processes.

> **Keyword definition**
>
> **Evolution** – The cumulative, gradual change in the genetic characteristics of successive generations of a species or race of an organism, ultimately giving rise to species or races different from the common ancestor. Evolution reflects changes in the genetic composition of a population over time.

■ Natural selection

Charles Darwin developed the **theory of evolution by natural selection** (Figure 3.4). This explained how the Earth's biodiversity has arisen:

- Populations show variation (i.e. not all individuals are the same).
- Populations always over-reproduce to produce excess offspring.
- Resources, such as food and space, are limited and there are not enough for all offspring.
- There is competition for resources.
- Due to variation within the species, some individuals will be fitter than others.
- Fitter individuals have an advantage and will reproduce more successfully than individuals who are less fit.
- The individuals that survive contain genes that give them an adaptive advantage.
- These genes are inherited by offspring and passed on to the next generation.
- Over time there is a change in the gene pool, which can lead to the formation of new species.

Figure 3.4 Charles Darwin published *The Origin of Species* in 1859; the book explained and provided evidence for the theory of evolution by natural selection

Speciation is the process by which new species form. Natural selection works with **isolating mechanisms** to produce new species (see below).

Expert tip

Natural selection contributes to the evolution of biodiversity over time.

Expert tip

Biological variation arises randomly and can either be beneficial to, damaging to, or have no impact on, the survival of the individual.

■ **QUICK CHECK QUESTIONS**

4   Define the term *evolution*.
5   Outline the theory of evolution by natural selection.

The role of isolation in forming new species

Revised ☐

■ Geographical isolation

Geographical isolation is caused by a physical barrier that leads to populations becoming separated, eventually leading to speciation (Figure 3.5). Causes of geographical isolation include plate activity (see page 70), and the formation of mountains, seas, lakes, rivers and deserts.

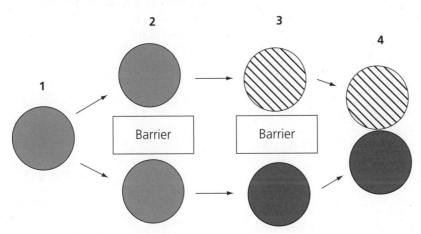

Figure 3.5 The process of speciation: 1 – original population; 2 – geographical isolation divides the population into two separate groups; 3 – populations become adapted to different local conditions and become genetically different from each other; 4 – reproductive isolation occurs and the species cannot interbreed to produce fertile offspring if they meet

■ Reproductive isolation

Reproductive isolation is caused by processes that prevent the members of two different species from producing offspring together. It includes:

- **environmental isolation** – the geographic ranges of two species overlap, but their niches differ enough to cause reproductive isolation
- **temporal isolation** – two species whose ranges overlap have different times of activity
- **behavioural isolation** – courtship rituals (breeding calls, mating dances etc.) between two species vary, such as in birds of paradise
- **mechanical isolation** – physical differences in, for example, reproductive organs, prevent mating or pollination
- **gametic isolation** – sperm and ova are incompatible, and will not allow fertilisation to take place

Keyword definitions

Isolation – The process by which two populations become separated by geographical, behavioural, genetic or reproductive factors. If gene flow between the two subpopulations is prevented, new species may evolve. See also **evolution**.

Speciation – The formation of new species when populations of a species become isolated and evolve differently from other populations. See also **evolution**.

Plate activity

The outer crust and upper mantle (the **lithosphere**) of the Earth are divided into many plates that move over the molten part of the mantle (**magma**).

- Plates move apart, slide against each other or collide.
- Plates move apart at **constructive plate margins** (Figure 3.6).
- Plates move together at **destructive plate margins** (Figure 3.7).
- Plates collide at **collision plate margins** (Figure 3.8).

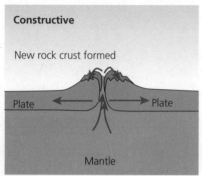

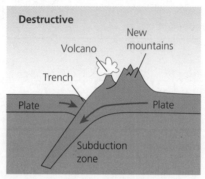

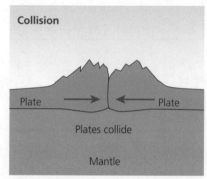

Figure 3.6 A constructive plate margin. Continental plates move apart (e.g. the East African rift valley system, including Lake Tanganyika and Lake Victoria). Magma rising from a gap in the crust may create new land (e.g. Iceland)

Figure 3.7 A destructive plate margin. Crust is **subducted** beneath (forced under) the other crust, causing rising magma to form new land (e.g. the Andes of South America)

Figure 3.8 A collision plate margin. Continental plates collide, leading to increased continental plate thickness and eventually new mountain ranges (e.g. the Himalayas, where the Indian plate is pushed against the Asian plate)

- The separation of continental plates leads to isolation of populations. This separates organisms with a common ancestor. Separation of gene pools results in divergent evolution, for example within the ratites (flightless birds): emu in Australia, ostrich in Africa, rhea in South America.
- Collision of plates can lead to uplift and mountain formation. The mountains form a physical barrier, which isolates populations. The uplift also creates new habitats, promoting biodiversity. Adaptation to new habitats then occurs through natural selection.
- Collision of plates can also cause the spread of species through the creation of land bridges. This leads to a mixing of gene pools and possible hybridisation.
- Plate activity can create new islands, usually through volcanic activity. This can lead to adaptations to fill new habitats/niches – see, for example, the Case study on page 71.
- The movement of plates to new climate regions leads to evolutionary change to adapt to new conditions, for example the northwards movement of the Australian plate.

Keyword definition

Plate tectonics – The movement of the eight major and several minor internally rigid plates of the Earth's lithosphere in relation to each other and to the partially mobile **asthenosphere** below.

Expert tip

You need to be able to explain how plate activity has influenced evolution and biodiversity.

GALÁPAGOS FINCHES

- The Galápagos Islands were created by rising magma from breaks in the crust ('hot spots').

- Volcanic islands were formed as a plate moved over the hot spots.

- An ancestral finch colonised the islands from mainland South America.

- Different populations of the finch became isolated on different islands.

- They adapted to the different conditions found on each island (see Figure 3.9).

- Galápagos finches have undergone speciation to fill many of the niches on these volcanic islands and they now are very different from the original mainland South American finch.

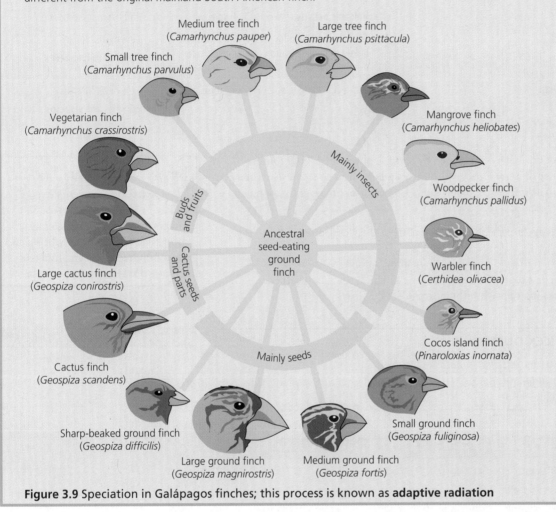

Figure 3.9 Speciation in Galápagos finches; this process is known as **adaptive radiation**

■ QUICK CHECK QUESTIONS

6 Outline how geographical isolation leads to speciation.

7 Explain the role of reproductive isolation in the formation of new species.

8 Outline the role of plate activity in speciation.

Environmental change gives new challenges to species: those that are suited will survive, and those that are not suited will not survive.

Past and present rates of species extinction

Revised

The fossil record shows that there have been five periods of mass extinction in the past (see Table 3.2). **Mass extinctions** include events in which 75% of the species on Earth disappear within a geologically short time period, usually between a few hundred thousand to a few million years.

Table 3.2 Past mass extinctions, during which 99% of all species that have ever existed on Earth have been lost

Period	Millions of years ago	Possible cause of mass extinction	Biodiversity loss
Cretaceous–Tertiary	65	Asteroid impact *or* Volcanic activity leading to climate change	16% of marine families 18% of land vertebrate families including dinosaurs **76% of all species**
End Triassic	199–214	Volcanic activity leading to climate change	23% of all families **80% of all species**
Permian–Triassic	251	Comet/asteroid impact *or* Volcanic activity reducing O_2 in sea	57% of all families **96% of all species** (largest mass extinction)
Late Devonian	364	Global cooling (followed by global warming), linked to diversification of land plants (less CO_2 in atmosphere)	19% of all families **75% of all species**
Ordovician–Silurian	439	Sea-level changes due to glacier formation	27% of all families **86% of all species**

Past mass extinctions have been caused by natural, physical (abiotic) causes. Scientists consider that the Earth is currently undergoing a **sixth mass extinction**, caused by human activities (i.e. biotic causes). The current mass extinction can be divided into two phases:

- Modern humans dispersed to different parts of the world, around 100 000 years ago.
- Humans started to grow food using agriculture, around 10 000 years ago.
- Scientists predict that at current rates of extinction the Earth will enter its sixth mass extinction within the next 300–2000 years.

> **Keyword definition**
>
> **Mass extinction** – events in which 75% of the species on Earth disappear within a geologically short time period, usually between a few hundred thousand to a few million years.

> ■ **QUICK CHECK QUESTIONS**
>
> **9** Define the term *mass extinction*.
> **10** What is the difference between past mass extinctions and the predicted sixth mass extinction?

> **Expert tip**
>
> You need to be able to discuss the causes of mass extinctions.

> **EXAM PRACTICE**
>
> **2 a** Define the term *species diversity*. [1]
>   **b** Explain how natural selection can produce new species. [2]
> **3** State, giving examples, *two* ways in which an understanding of plate tectonics has helped to explain patterns of biodiversity. [4]

3.3 Threats to biodiversity

Revised ☐

> **SIGNIFICANT IDEAS**
>
> - While global biodiversity is difficult to quantify, it is decreasing rapidly due to human activity. Classification of species conservation status can provide a useful tool in the conservation of biodiversity.

The number of species on Earth

Scientists currently do not know how many species exist on Earth. There are around 1.8 million described species, but scientists can only estimate how many other organisms remain undiscovered. The number of species is poorly known for many reasons:

- Some habitats are difficult to reach, such as the canopy of tropical rainforest, and may contain many undescribed species.
- Lack of finance for scientific research has resulted in many habitats and groups being significantly under-recorded.
- There are difficulties with classification. Description of new species in the past has been based on physical characteristics, which may be less accurate at distinguishing species than modern DNA techniques.
- Bias in the past led to the discovery of large-sized popular and appealing groups such as mammals and birds, which represent a very small proportion of total species numbers, and so smaller species requiring microscopic examination are less well known.
- Only a relatively small number of taxonomists globally are involved in identifying and describing new species.

Estimates of the total number of species on Earth vary considerably from 5–100 million. Higher numbers may be possible if groups such as bacteria and soil organisms are included, which are currently poorly understood. Estimates are based on mathematical models, which depend on the accuracy of the data collected.

The need to understand more about the range and diversity of life has become increasingly important. Estimations of species extinction rates need to based on known species numbers, so that a percentage loss of different groups can be calculated.

Expert tip

The total number of classified species is a small fraction of the estimated total of species, and it continues to rise. Estimates of extinction rates as a consequence are also varied, but current extinction rates are thought to be between 100 and 10 000 times greater than background rates.

Factors that lead to loss of diversity

Natural events can cause a loss of diversity. Examples include:

- volcanic activity
- drought/floods
- ice ages
- meteor impacts.

Human actions can cause a loss of diversity. Examples include:

- agricultural practices such as **monoculture** (a crop of only one species), use of **pesticides** and use of **genetically modified species**
- habitat degradation, fragmentation and loss
- introduction of non-native (**invasive**) species
- pollution
- population growth, leading to disturbance of habitats, pollution, etc.
- **overhunting**, collecting and harvesting.

The rate of biodiversity loss will vary from country to country depending on:

- the ecosystems present
- the protection policies and monitoring systems in place
- EVSs of the local residents
- the stage of economic development (i.e. LEDC or MEDC).

Expert tip

A way of remembering the human causes of biodiversity loss is the mnemonic 'A HIPPO': **A**griculture, **H**abitat loss, **I**nvasive species, **P**ollution, **P**opulation (i.e. the effects of population growth) and **O**verhunting.

■ QUICK CHECK QUESTIONS

11 List *three* natural causes of biodiversity loss.

12 List *six* human causes of biodiversity loss.

13 Outline how agricultural practices have led to a loss of biodiversity.

Tropical biomes and sustainable development

Tropical biomes, such as tropical rainforest and mangrove (Chapter 2, pages 47–48) contain some of the most globally biodiverse areas. The unsustainable exploitation of tropical biomes (see Case study, below) results in massive losses in biodiversity and their ability to perform globally important ecological services such as water and carbon cycling.

Most tropical biomes occur in less economically developed countries (LEDCs) and therefore there is conflict between exploitation, sustainable development and conservation:

- Clearance of ecosystems provides land for cash-crops such as oil palm.
- Plantations provide financial income for local communities.
- Clearance leads to biodiversity loss.

For sustainable development in LEDCs, there needs to be a balance between conserving tropical biomes and using the land to provide income for the local economy.

Diversification of the local economy into areas such as ecotourism can provide alternative sources of income and allows conservation areas to be established, protecting biodiversity.

> **Expert tip**
>
> You need to be able to evaluate the impact of human activity on the biodiversity of tropical biomes. You also need to be able to *discuss* the conflict between exploitation, sustainable development and conservation in tropical biomes.

CASE STUDY

VULNERABILITY OF TROPICAL RAINFOREST

Tropical rainforests cover only 5.9% of the Earth's land surface but may contain up to 50% of all species. They are found in South America, Africa and Southeast Asia.

The climate is warm and stable:

- Temperatures vary from 20°C at night to 35°C at midday.
- Rainfall is high, with up to 2500 mm per year.

The constant warm temperatures, high insolation and high rainfall lead to high levels of photosynthesis and high productivity (net primary productivity, NPP – see page 41):

- High productivity leads to high biomass.
- This leads to ecosystem complexity, abundant resources (e.g. food) and niche diversity.
- Abundant niches lead to high species richness.
- Biodiversity is therefore high (rainforests contain **biological hotspots**).

Tropical rainforests are vulnerable to disturbance, with an average of 1.5 hectares (equivalent to a football pitch) lost every 4 seconds. Because they have high biodiversity, many species are affected when they are disturbed. Deforestation and forest degradation are being caused by demands for:

- timber
- land for cattle to provide beef
- soya and biofuels (e.g. oil palm in Southeast Asia).

Tropical rainforests are found on nutrient-poor soils that are thin and easily eroded once forest is cleared. The time taken for rainforest to regenerate depends on the level of disturbance:

- Small-scale disturbance, for example from shifting cultivation (pages 110–112), can recover in around 50 years.
- Large areas of cleared land will take longer to grow back (around 4000 years), if at all.
- Areas subject to selective logging methods will regenerate more quickly than areas logged using conventional methods.

Threats to tropical rainforest in the 1970s and 1980s led to the growth of the Green movement. **Green politics** is an ideology that places central importance on ecological and environmental goals, and sustainable development. The Green movement aims to reduce deforestation and increase reforestation.

Factors that make species prone to extinction

Not all species are equally vulnerable to extinction. Certain animals and plants, through their ecology or behaviour, are more at risk. Factors include:

- small population size, which leads to a reduced gene pool and therefore the species is more prone to disease and inbreeding, and susceptible to environmental change (e.g. Asiatic cheetah)
- limited distribution (e.g. golden lion tamarin monkey – see page 78)
- high degree of specialisation in, for example, dietary needs (e.g. giant panda, which mainly eats bamboo)
- slow reproductive rate, i.e. *K*-selected species with a small number of young (e.g. western lowland gorilla)
- low reproductive potential (e.g. Bicknell's thrush in Canada only has 2000–5000 breeding pairs)
- non-competitive/altruistic behaviour (e.g. the dodo, extinct since the late seventeenth century)
- high trophic level, such as a top carnivore, which can accumulate toxins (e.g. Hawaiian monk seal)
- long migration routes (e.g. Siberian crane)
- complex migration routes (e.g. southern bluefin tuna)
- habitat under threat (e.g. Sumatran tiger – see page 77)
- human pressure from hunting, collecting, trade etc. (e.g. Sumatran rhino).

> **Expert tip**
>
> Natural factors that lead to a loss of biodiversity operate at two different orders of magnitude:
>
> - Extinction at a local level (**background extinction**) caused by, for example, droughts, floods, habitat loss, disease or the evolution of a superior competitor.
> - **Mass extinction** caused by global catastrophic events, for example volcanic activity, meteor impact and glacial events causing changes in sea level.

■ Determining a species' Red List conservation status

The **International Union for Conservation of Nature (IUCN)** has worked for more than four decades to assess the conservation status of species on a global scale. The IUCN presents this information in the **Red List**. This is done to:

- highlight species threatened with extinction
- promote conservation of threatened species.

Different factors are used to determine a species' conservation status. A sliding scale operates, from *least concern* to *extinct*. Factors include:

- population size (smaller populations are more likely to go extinct)
- reduction in population size (indicating that a species is under threat)
- degree of specialisation
- distribution
- geographic range and degree of fragmentation
- degree of endemicity (i.e. only found in one specific area)
- quality of habitat (species are less likely to survive in poor habitats)
- trophic level (animals in higher trophic levels are more likely to go extinct)
- probability of extinction.

■ QUICK CHECK QUESTIONS

14 State *three* characteristics that might make an organism vulnerable to extinction.

15 Distinguish between the terms *background extinction* and *mass extinction*.

16 State the role of the IUCN Red List.

17 Outline the factors that are used to determine Red List conservation status.

Common mistake

If an exam question asks you to outline factors that make a species prone to extinction use examples such as those given here: do not talk about general causes for the loss of biodiversity, for example floods, droughts and volcanic activity.

Extinct, critically endangered and back from the brink

Revised ☐

CASE STUDY

THE ELEPHANT BIRD OF MADAGASCAR

Figure 3.10 An elephant bird egg compared with a chicken egg

Table 3.3 Extinction of the elephant bird

Species	Elephant bird
Conservation status	Extinct
Threats to species	Endemic to the island of Madagascar and therefore, once lost there, it was lost globally. Humans arrived on the island from Southeast Asia, which led to the extinction of the giant bird around 1000 years ago: ● The forest the species relied on for survival was cleared and burned. ● Around 80% of the original natural vegetation was lost. ● The habitat was cleared to create land for farming (rice and grazing land for cattle). ● The surviving forest did not have enough vegetation to support the giant bird, or it included vegetation of the wrong sort (e.g. baobab trees). ● The elephant bird's nutritious eggs were eaten by humans (one egg ≈ 140 chicken eggs – the largest egg of any species that has existed).
Ecological role	Herbivore
Effects on ecosystem when the species disappeared	Spiny seeds of the endemic plant *Uncarina* were dispersed on the feet of the elephant bird. The loss of the elephant bird would have affected the dispersal of the plant, although domestic animals introduced to the island may now help to disperse the seeds. The forest coconut is a critically endangered endemic species that may have been adapted for passage through the elephant bird gut – the current poor dispersal of this species may have resulted from the extinction of the elephant bird. The species was the heaviest bird that has existed (around 0.5 tonne) and would have eaten large amounts of vegetation, for example in the spiny forest to the south of the island. The loss of the species would have left more vegetation for other species such as sifaka lemurs.

CASE STUDY

THE SUMATRAN TIGER

Figure 3.11 A Sumatran tiger

Table 3.4 The Sumatran tiger – a species currently critically endangered and heading towards extinction

Species	Sumatran tiger
Conservation status	Critically endangered
Threats to species	Has a small and declining population for the following reasons: ● Loss of habitat (tropical rainforest). ● It is seen as a danger to humans and livestock and so is hunted. ● Fragmentation of its habitat makes breeding difficult. ● The high market value of its body parts encourages poaching. ● As a top predator its population is small because little energy reaches the top of the pyramid. ● The species is only found on one island (Sumatra) and so it is prone to extinction. ● A large area is needed to maintain a viable population, but tropical rainforest on the island of Sumatra is rapidly being cut down. ● The small population size leads to low genetic diversity, which leaves the species more prone to disease.
Ecological role	Top carnivore
Possible effects on ecosystem if the species disappeared	Species at the trophic level below would become more numerous. The shortened food chain produces imbalances at other trophic levels. Sick or weak animals lower down the food chain, usually eaten by this species, are no longer killed. Less fit individuals lower down the food chain survive to breed. Decomposer organisms associated with the tiger's dung are lost.

Expert tip

You need to be able to discuss the case histories of three different species: one that has become extinct due to human activity, another that is critically endangered, and a third species whose conservation status has been improved by intervention.

THE GOLDEN LION TAMARIN MONKEY

Figure 3.12 A golden lion tamarin monkey

Table 3.5 The golden lion tamarin monkey – a species currently critically endangered but whose conservation status has been improved by intervention

Species	Golden lion tamarin monkey
Conservation status	Critically endangered, but conservation status has been improved by intervention
Threats to species	Under threat for the following reasons: • 90% of their original forest habitat (tropical rainforest) in Brazil has been cut down. • The remaining habitat is small and fragmented. • The species is only found in one small area of Brazil, and is therefore especially prone to extinction. • Thought by some people to be carriers of human diseases such as yellow fever and malaria, and were killed for this reason. • Kept as pets as part of the exotic pet trade.
Ecological role	Omnivore
Possible effects on ecosystem if the species disappeared	The golden lion tamarin eats fruits, insects and small lizards. Loss of the species would mean the following: • Seed dispersal of plants that have fruits eaten by the monkey would be affected. • Species at lower trophic levels, such as insects and small lizards, would become more numerous. • Species at higher trophic levels would become less numerous. • Shortened food chains would produce changes in other trophic levels and produce imbalances in the forest food web.
How are they being restored?	Numbers in the wild have increased from a low of 400 in the 1970s to around 1000 today. Captive breeding programmes in zoos (*ex situ* **conservation**) have produced numbers that allow release into the wild. There are efforts to preserve native forests for *in situ* **conservation**, for example Reserva Biologica de Poyo das Antas, near Rio de Janeiro.

Natural area of biological significance under threat

Revised ☐

Expert tip

You need to be able to describe the threats to biodiversity from human activity in a given natural area of biological significance or conservation area.

CASE STUDY

BORNEO'S RAINFOREST

Table 3.6 The threats faced by Borneo's rainforest

Ecosystem	Borneo tropical rainforest
Description	<ul><li>Borneo is the third largest island in the world and has historically been covered by tropical rainforest.</li><li>Rainforest is dominated by dipterocarp trees – long-lived, tall, hardwood trees that are valuable timber species.</li><li>The island has high species diversity, with 15 000 plant species, 220 mammal species and 420 bird species.</li><li>Around 20% of the mammal species are endemic to Borneo.</li><li>300 species of tree can be found in 1 hectare of forest.</li><li>Many species are on the Red List – for example orang-utan, sun bear, Asian elephant and Sumatran rhino.</li></ul>
Natural threats	<ul><li>Natural forest fires</li><li>Drought</li><li>Dry periods (El Niño) are caused by the Southern Oscillation, where surface air pressure and ocean temperatures fluctuate between the eastern and western tropical Pacific, causing drier periods in one location and wetter periods in the other.</li><li>El Niño contributed to widespread fires in 1982–1983 and 1997–1998.</li><li>Borneo lies south of the Pacific hurricane belt and so is not subject to these very high winds, although tropical storms can cause canopy or emergent trees to fall, with many neighbouring trees, attached by lianas, being brought down with them.</li></ul>
Human threats	<ul><li>The rainforest has been commercially logged since the 1970s for export markets, with logging accelerating in the 1980s and 1990s.</li><li>In 1974 forest cover was estimated to be 6.4 million ha (88% of the total land area) whereas only 4.5 million ha remained in 1985 (i.e. a 30% reduction in 11 years).</li><li>The deforestation rate (2000–2005) was 3.9% per annum.</li><li>At the peak of logging operations, trees were removed in large numbers (up to 100 m³ ha⁻¹ in volume).</li><li>Conventional logging methods were not selective and caused damage to the remaining forest.</li><li>More recently, selective logging (reduced-impact logging – RIL) methods have been used. These cause less damage and allow faster regeneration of forest.</li><li>Since the 1990s oil palm has been planted on cleared land (see Figure 3.13).</li><li>One hectare of oil palm yields up to 5000 kilograms of crude palm oil.</li><li>Most oil palm plantations are owned by the state or by transnational corporations.</li></ul>
Consequences of disturbance	<ul><li>Damage to the remaining forest is proportional to the amount of timber removed.</li><li>Heavily logged forest using conventional methods can remove 94% more timber than in RIL sites.</li><li>Conventional logging methods can severely damage forest structure, leading to an increase in light-loving, riverine species such as *Macaranga* trees.</li><li>Changes in forest structure reduce biodiversity.</li><li>RIL is better at preserving forest structure and biodiversity.</li><li>Oil palm plantations are monocultures with low species diversity.</li><li>Oil palm plantations fragment rainforest, block migration routes and remove habitats for animals.</li><li>Insecticides and herbicides are used to control insect pests and weeds, and so reduce biodiversity.</li><li>Animals such as Asian elephants and orang-utans that stray into the plantations can be illegally killed.</li><li>Without forest cover, soil is eroded, making it difficult/impossible for the original vegetation to regrow.</li><li>Loss of forest cover causes changes to stream flow and reduces stream diversity.</li><li>Loss of forest reduces transpiration from leaves, which affects local weather patterns, leading to drier areas more prone to fire.</li><li>The valuable role that the ecosystem provides, through biodiversity and controlling weather patterns, has been reduced.</li><li>The role of the rainforest as an economic resource (e.g. though ecotourism) has been reduced.</li></ul>

Figure 3.13 Palm oil is used as a food (e.g. in margarine, cooking oil, ice cream, ready meals, biscuits and cakes) and is also used in the production of detergents and cosmetics. Oil palm has replaced tropical rainforest over large areas of Borneo

■ QUICK CHECK QUESTIONS

18 Describe the ecological role of a named extinct species and outline the possible consequences of its disappearance.

19 With reference to the case history of a named critically endangered species describe the human factors that have led to its conservation status.

20 Describe how the conservation status of a named endangered species is being improved by intervention.

21 Describe the case history of a natural area of biological significance that is threatened by human activities.

EXAM PRACTICE

4 Outline three reasons why tropical rainforests are vulnerable to habitat destruction. [3]

5 Name a species of plant or animal that has become extinct, and list two factors that help to explain why that species became extinct. [2]

6 List three characteristics that might make some bird species more prone to extinction than others. [2]

7 Suggest why more extinctions can be expected on islands than on continents. [5]

8 With reference to a named ecosystem describe the natural and human threats it faces and discuss the consequences of this disturbance. [10]

Expert tip

You need to be able to describe the ecological, socio-political and economic pressures that have caused, or are causing, the degradation of an area of biological significance, and the consequent threat to biodiversity.

3.4 Conservation of biodiversity

Revised ☐

SIGNIFICANT IDEAS
- The impact of losing biodiversity drives conservation efforts.
- The variety of arguments given for the conservation of biodiversity will depend on EVSs.
- There are various approaches to the conservation of biodiversity, each with associated strengths and limitations.

Arguments for preserving species and habitats

There are many different arguments for preserving biodiversity. These include:

- **ethical** reasons (e.g. every species has a right to survive; we have a responsibility to safeguard resources for future generations) – these are very broad, and can include the intrinsic value of the species or the utilitarian value
- **aesthetic** (i.e. visual) reasons (e.g. provides beauty and inspiration)
- **economic** reasons:
  - ☐ value of genetic resources for humans (e.g. improved crops)
  - ☐ commercial considerations of the natural capital (e.g. new medicines)
  - ☐ value of ecotourism (which benefits from higher levels of diversity)
- **ecological** reasons:
  - ☐ conserving rare habitats (e.g. endemic species require specific habitats)
  - ☐ ecosystems with high levels of diversity are generally more stable
  - ☐ healthy ecosystems are more likely to provide ecological services (e.g. pollination; flood prevention)
  - ☐ species diversity should be preserved as it can have knock-on effects on the rest of the food chain
- **social** reasons:
  - ☐ loss of natural ecosystems can lead to loss of peoples' homes, sources of livelihood and culture
  - ☐ areas of high biodiversity provide income for local people through, for example, tourism, and so support social cohesion and cultural services.

Ecosystems provide a variety of different goods and services:

- **Goods,** for example food, fibre, fuel (peat, wood and non-woody biomass) and water from aquifers, rivers and lakes. Goods can also be from heavily managed ecosystems (intensive farms and fish farms) or from semi-natural ones (such as by hunting and fishing).
- **Support services** – essentials for life, including primary productivity, soil formation and the cycling of nutrients.
- **Regulatory services,** for example pollination, regulation of pests and diseases, climate regulation, flood regulation, water quality regulation and erosion control.
- **Cultural services,** providing opportunities for outdoor recreation, education, spiritual wellbeing and improvements to human health.

It is easier to give value to some aspects of biodiversity than others. For example, commercial (i.e. economic) value can easily be applied to goods such as timber, medicine and food. It is more difficult to give value to such things as ecosystem support services, ecosystem regulatory services, cultural services, and ethical and aesthetic factors.

The role of intergovernmental and non-governmental organisations

- **Intergovernmental organisations (IGOs)** are established through international agreements. They bring governments together to work to protect the Earth's natural resources. An example is the United Nations Environment Programme (UNEP).
- **Non-governmental organisations (NGOs)** are not run by, funded by, or influenced by governments of any country. Examples include Greenpeace and the World Wildlife Fund (WWF).

Both governmental organisations and non-governmental organisations are involved in preserving and restoring ecosystems and biodiversity. They have contrasting roles and activities.

Table 3.7 Contrasting roles and activities of NGOs and IGOs

	NGO	IGO
Speed of response	Rapid: organisations can make their own decisions	Slow: there must be agreement between governments
Use of media	Use film of activities (e.g. chasing whaling boats) to gain media attention	Professional media liaison officers prepare and read written statements
Diplomatic constraints	Unaffected by political considerations Activities can be illegal, although this is discouraged	Many constraints: cannot make decisions without agreement from all parties Disagreements can cause serious constraints
Political influence	Green politics can establish environmental issues as part of the political process	Direct access to the governments of many countries
Enforceability	No direct power: must use public opinion to persuade governments to act	Use international treaties and national or state laws to protect the environment, ecosystems and biodiversity

International conventions on biodiversity

Revised

International conventions have shaped attitudes towards sustainability. The **UN Conference on the Human Environment** (Stockholm, 1972) was the first time that the international community met to consider global environment and development needs together. It led to the **Stockholm Declaration**, which played an essential role in setting targets and triggering action at both local and international levels.

In 1992 the UN **Rio Earth Summit** resulted in the **Rio Declaration** and **Agenda 21**:

- The Earth Summit was attended by 172 governments, and set the agenda for the sustainable development of the Earth's resources.
- The Earth Summit led to agreement on two legally binding conventions: the **UN Convention on Biological Diversity (UNCBD)** and the **UN Framework Convention on Climate Change (UNFCCC)**.
- Both the CBD and FCCC are governed by the **Conference of the Parties (CoP)**, which meets either annually or biennially to assess the success and future directions of the Convention. For example CoP 11 of the CBD took place in India in 2012, and CoP 18 of the FCCC took place in Doha in 2012.
- In 1997 a follow-up meeting (Rio +5) took place in New York to assess the succession of the Earth Summit and future directions. A number of gaps were identified, for example in social equity and poverty.
- Ten years after Rio, the **Johannesburg World Summit on Sustainable Development** addressed mainly social issues, with targets set to reduce poverty and increase access to safe drinking water and sanitation.
- In 2012, Rio +20 took place to mark the twentieth anniversary of the Earth Summit. It had three main objectives: to secure political commitment from nations to sustainable development; to assess progress towards internationally agreed commitments (e.g. CO_2 reductions); and to examine new and emerging challenges.

In 1997, the **Kyoto Protocol** built on the Earth Summit's Climate Change Convention. It was developed at a UN meeting in Kyoto, and agreed to reduce greenhouse gas emissions from 1990 levels.

Expert tip

You need to know about recent international conventions on biodiversity, such as those signed at the Rio Earth Summit (1992) and subsequent updates.

■ The World Conservation Strategy

The **World Conservation Strategy (WCS)** was established in 1980 by the International Union for Nature Conservation (IUCN). The IUCN is concerned with the conservation of resources for sustainable economic development. The WCS consisted of three aims:

- maintaining essential life-support systems (climate, water cycle, soils) and ecological processes
- preserving genetic diversity
- using species and ecosystems in a sustainable way.

It outlined a series of global priorities for action, while recommending that each country prepare its own national strategy as a developing plan that would take into account the conservation of natural resources for long-term human welfare.

The WCS emphasised the importance of making the users of natural resources become their guardians and recognised that conservation plans can only succeed with the support and understanding of local communities. It focused on specific arguments for preserving biodiversity (i.e. socio-economic, genetic resource and ecological arguments) because:

- these are more universally agreed by people with different environmental viewpoints
- ethical and aesthetic arguments are more difficult to define and can vary between different communities
- these arguments are more scientifically verifiable than ethical or aesthetic arguments
- most influential nations, and those nations involved in drawing up the WCS, attach more value to scientific validity than other arguments.

■ QUICK CHECK QUESTIONS

22 Outline *five* different arguments for preserving biodiversity.

23 Suggest why it is easier to give commercial value to some aspects of biodiversity than others.

24 Compare the roles and activities of governmental organisations and non-governmental organisations.

Designing a protected area

Revised ▢

Protected areas should aim to preserve the greatest amount of natural habitat within an ecosystem, and therefore maintain the complex ecological interactions that maintain equilibrium and biodiversity. In most countries, protected areas are islands surrounded by areas of disturbance.

Island biogeography theory predicts that smaller islands of habitat will contain fewer species than larger islands. It is therefore inevitable that protected areas will have lost some of the diversity seen in the original undisturbed ecosystem. The principles of island biogeography can be applied to the design of reserves (Figure 3.14).

Table 3.8 Features of protected areas that are better or worse for conservation

	Better	Worse
A	Single large area	Single small area
B	Single large area	Several small areas of the same total size
C	Intact habitat	Fragmented and disturbed habitat
D	Areas connected by corridors	Separated areas
E	Round (= fewer edge effects)	Not round (= more edge effects)

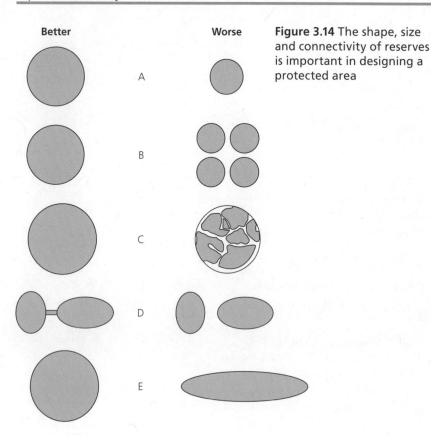

Figure 3.14 The shape, size and connectivity of reserves is important in designing a protected area

Protected areas that are better for conservation have the following features:

- Larger and so support a greater range of habitats, and therefore greater species diversity.
- Higher population numbers of each species.
- Greater productivity at each trophic level, leading to longer food chains and greater stability.
- Maintain a low perimeter:area ratio to reduce edge effects. Fewer edge effects mean more of the area is undisturbed.
- Maintain top carnivores and large mammals by having a large area.
- If areas are divided, then fragmented areas need to be in close proximity to allow animals and plants to move between them.
- Maintain gene flow between fragmented reserves by allowing movement along corridors.
- Allow movement of large mammals and top carnivores between fragments by maintaining corridors.

Buffer zones

Buffer zones help to protect conservation areas and maintain equilibrium and biodiversity. They contain habitats that are either managed or undisturbed, and minimise disturbance in the protected area from outside influences such as people, agriculture, pests and diseases.

In Figure 3.15, for example, salak palm provides a barrier to stop illegal hunting, encroachment of farmers and illegal felling of trees in the protected area. The salak palm also provides a barrier to stop animals leaving the reserve and being killed in the surrounding area.

The sugar cane is grown by local farmers to provide a cash crop, while fruits from the salak palm also provide food. The crops provide protection against fire.

Expert tip

You need to be able to discuss the best design for protected areas using the criteria of size, shape, edge effects, corridors and proximity.

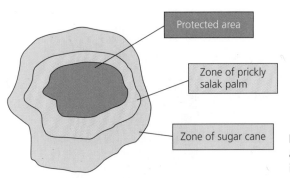

Figure 3.15 Buffer zones around protected forest in Borneo

■ Evaluating the success of a protected area

Successful protected areas have the following characteristics:

■ Provide vital habitat for indigenous species. This can include habitat and food for migrating species such as birds.
■ Create community support for the area.
■ Receive adequate funding and resources.
■ Carry out relevant ecological research and monitoring.
■ Play an important role in education.
■ Protected by legislation.
■ Have policing and guarding policies.
■ Give the site economic value.

> **Expert tip**
>
> You need to be able to explain the criteria used to design and manage protected areas.

CASE STUDY

THE KABILI-SEPILOK FOREST RESERVE, BORNEO

The Kabili-Sepilok forest reserve is a 4300 ha area of lowland rainforest in Sabah, Malaysian Borneo. The forest is rich in dipterocarp trees (hardwood species) and contains a wealth of natural wildlife, including orang-utan, leaf monkeys and gibbon. The success of the conservation area is due to many different factors:

Local support

■ Local guides and rangers earn a living assisting tourists within the park and protect the forest and its biodiversity.
■ There are many ecotourism resorts near the reserve run by local people.
■ Their economic future and the future of the park are closely linked.
■ They have a positive vested interest in the conservation area.
■ They have a respect for, and pride in, the reserve.

Government agencies

■ The Sabah Wildlife Department and Forestry Department help to manage the reserve.
■ The government, through its employees, wildlife agencies, rangers and guides, provides the park with security and infrastructure.
■ These government agencies monitor and control visitor numbers and help to protect the reserve.
■ They provide resources.
■ They liaise with local groups, non-governmental groups and international organisations.

Research

■ The Forest Research Centre (FRC) at Sepilok carries out scientific research within the reserve:
■ This allows its ecosystems and biodiversity to be monitored.
■ It allows new information to be discovered.
■ Research identifies new hazards and new goals.
■ The FRC produces information that supports the park's existence and informs management decisions.

■ It helps educate those inside and outside the park, both nationally and internationally.

■ Research is also carried out by the orang-utan rehabilitation centre at the edge of the reserve (Figure 3.16; see below), which returns captive animals to the wild.

Education

■ The orang-utan rehabilitation centre rehabilitates orphaned orang-utans. It is a major international tourist attraction, focusing on public education, research and conservation.

■ A Rainforest Discovery Centre (RDC) provides environmental education facilities for students, teachers and overseas tourists.

■ The Bornean Sun Bear Conservation Centre (BSBCC) was created to rehabilitate captive sun bears back into the wild. The BSBCC promotes greater awareness of the ecology of the bear and the threats it faces.

Figure 3.16 Orang-utan at the rehabilitation centre, Sepilok

A holistic approach to conservation

■ Sepilok is an example of a **holistic approach to conservation**: it is not just an area of wildlife protection, but also one where educational activities are encouraged, research takes place, people use it as an area of relaxation and its cultural value is encouraged (e.g. the national importance of the wildlife to Malaysia).

■ Protection without considering other factors, for example economics, culture and development, is unlikely to be successful.

■ Multiple-use reserves are more popular and easier to fund, and are more sustainable in the long term.

Limitations

Although the impact of Sepilok has been overwhelmingly positive, there are some limitations.

■ The Sepilok Forest Reserve is a relatively small area surrounded by developed areas such as oil palm plantations. The forest is not large enough to sustain populations of large mammals and important species such as the orang-utan. Populations of orang-utan are supported by feeding from the rehabilitation centre.

■ A large number of tourists who visit the centre visit for a morning or afternoon and do not stay in the area, so local communities do not benefit financially from the increased number of visitors.

■ Whilst the internationally famous Sepilok forest reserve is maintained, other areas that are equally biologically significant are not conserved and are degraded or cleared to make room for plantation forest.

Overall, however, the impacts of Sepilok have been overwhelmingly positive, drawing many tourists to the area and raising the profile of Malaysian conservation internationally.

Expert tip

The granting of protected status to a species or ecosystem is no guarantee of protection without community support, adequate funding and proper research.

Expert tip

You need to be able to evaluate the success of a given protected area.

Expert tip

The location of a conservation area in a country is a significant factor in the success of the conservation effort. Surrounding land use for the conservation area and distance from urban centres are important factors for consideration in conservation area design.

Strengths and weaknesses of the species-based approach to conservation

Revised

The **species-based approach to conservation** focuses on vulnerable species and in raising their profile. It attracts attention and therefore funding for conservation, and can successfully preserve species in zoos and botanic gardens. There is a tendency to focus on the conservation of high-profile, charismatic species that catch public attention both nationally and internationally.

Species-based conservation involves:

- keystone species
- flagship species
- the Convention on International Trade in Endangered Species (CITES)
- captive breeding and reintroduction programmes
- zoos.

Keystone species

Keystone species are essential for the functioning of the ecosystem and in protecting the integrity of a food web.

In tropical rainforest, fig trees are a keystone species. The figs they produce are fed on by birds, such as hornbills (Figure 3.17), and mammals, such as organ-utan, and these animals rely heavily on this resource during the times of the year when other food is uncommon. Without figs, many species would disappear from the community.

Figure 3.17 A rhinoceros hornbill feeding on figs in the canopy of tropical rainforest in Malaysia. Figs are keystone species in these forests

Flagship species

The selection of 'charismatic' species can help to protect others in an area. Such species are known as **flagship species**. The Bengal tiger is an example of a flagship species (see also page 89).

CITES

CITES is the **Convention on the International Trade in Endangered Species (of Wild Fauna and Flora)**. It is an international agreement aimed at preventing trade in endangered species of plants and animals, and therefore:

- reduces demand for trade
- contributes to species conservation.

Under the convention countries agree to monitor trade in threatened species (and their products) at ports and airports. Illegal imports and exports are confiscated, which discourages illegal trade. If trade in whole organisms or parts of organisms can be reduced, pressure on wild populations is reduced.

CITES has helped to protect elephants and rhinos by reducing trade in ivory and rhino horn.

Table 3.9 Strengths and limitations of CITES

Strengths	Limitations
It is supported by many countries (145).	Enforcement is difficult.
It lists many species (around 700).	Fines are relatively small and may not deter poaching.
It bans commercialisation of many products and species.	Many countries have not signed and so are not subject to the agreement.
It has proved to be successful for many species, including elephants, rhinos, marine turtles, tigers, parrots and snakes.	Support by some countries is limited and ineffectual.
	The treaty favours large, conspicuous, attractive organisms.
	Despite the agreement, illegal hunting still occurs, including the poaching of ivory in Africa.

■ Captive breeding and reintroduction programmes

Captive breeding and **reintroduction programmes** are part of a species-based approach to conservation:

- They are usually done by zoos.
- A small population is obtained from the wild or from other zoos.
- Enclosures for animals are made as similar to the natural habitat as possible.
- Breeding can be assisted through artificial insemination.

Table 3.10 Strengths and limitations of captive breeding and reintroduction programmes

Strengths	Limitations
• Populations can build up quickly as habitat and food are abundant.	• Do not directly conserve the natural habitat diversity of the species. Conservation of habitat should lead to conservation of species.
• Abundant food and habitat reduces competition.	• Not all species breed easily in captivity (e.g. giant pandas).
• Allow predators and diseases to be controlled.	• It is difficult to maintain genetic diversity and so gene pools of species may be small.
• Individual animals can be exchanged between collections to prevent inbreeding and to maintain genetic diversity.	• Released animals may be easy targets for predators.
• Successful examples include the Arabian oryx and golden lion tamarin monkey (see page 78).	• Aesthetic values can lead to an imbalance in conservation activity, meaning that popular, charismatic species are conserved (e.g. Madagascan lemurs) while small, less-popular animals may not be part of the conservation programme (e.g. endemic Madagascan hissing cockroaches).
	• Some countries may have technical or economic difficulties in establishing programmes.

■ Zoos

Zoos protect species in carefully controlled environments. They are an example of *ex situ* conservation.

Table 3.11 Strengths and limitations of zoos

Strengths	Limitations
• They allow education through visits and so the public are more likely to support conservation campaigns.	• There are ethical arguments about keeping animals in captivity for profit.
• Genetic monitoring can take place.	• If the zoo is not properly managed, poor conditions can lead to psychological and physiological problems with animals.
• They allow captive breeding and reintroduction programmes (see above).	• Zoo animals may be unable to adapt back to life in the wild.
• The number of offspring surviving to adulthood is higher, so species numbers increase more rapidly.	• They can focus on high-profile/charismatic species and so can be less successful at saving 'non-cuddly' species.
• Studying species in zoos allows a better understanding of these animals, leading to improved management of the species outside zoos.	• Saving a species should require preserving the animal's habitat, which also benefits all other species.
• They can be used as an 'ark', preserving a species until its habitats are protected or restored.	• A species can be artificially preserved in a zoo while its natural habitat is destroyed (e.g. giant panda).

■ QUICK CHECK QUESTIONS

25 Describe and explain the criteria needed to design a successful protected area.

26 Evaluate the success of a named protected area.

27 Discuss the strengths and weaknesses of a species-based approach to conservation.

Comparing different approaches to conservation: summary

Table 3.12 Strengths and limitations of different approaches to conservation

	Strengths	Limitations
CITES	Can protect many species Signed by many countries Treaty works across borders	Not legally binding Difficult to enforce How it is implemented varies from country to country
Protected areas	Can conserve whole ecosystems Allow research and education Preserve many habitats and species Prevent hunting and other disturbance from humans Allow for *in situ* conservation	Can be very expensive Difficult to manage Subject to outside forces that are difficult to control Difficult to establish in the first place due to political issues/vested interests
Zoos	Allow controlled breeding and maintenance of genetic diversity Allow research Allow for education Effective protection for individuals and species	Have historically preferred popular animals; not necessarily those most at risk Problem of reintroducing zoo animals to wild *Ex situ* conservation and so do not preserve native habitat of animals Limits freedom of animals

A mixed approach to conservation

Revised ☐

A mixed approach to conservation means combining both *in situ* (e.g. protected areas) and *ex situ* (e.g. zoos and captive breeding) methods. This approach can be the best solution for species conservation in many instances.

CASE STUDY

THE BENGAL TIGER

The Bengal tiger is found mainly in India, with smaller populations in Bangladesh, Nepal, Bhutan and China. Although it is one of the most numerous tiger species, with more than 2500 left in the wild, the species is under threat from habitat loss and poaching. Tiger reserves established throughout India in the 1970s helped to stabilise numbers and are helping to protect the species.

Mangrove forests (see page 48) in Bangladesh and India, known as the Sundarbans, are the only mangrove forests where tigers are found. The Sundarbans are threatened by increasing sea level caused by climate change. As a flagship species (page 87), the Bengal tiger is helping to attract tourists to the Sunderbans and other areas, helping to raise money for conservation.

Figure 3.18 The Bengal tiger – an example of a flagship species

'Project Tiger' was launched in India in 1972, with the aim of both preserving areas where the tiger is found as well as ensuring the ongoing survival of a healthy population of tigers. Captive breeding of Bengal tigers is helping to maintain the genetic diversity of the species. A Bengal tiger studbook keeps a record of which Bengal tigers are kept in captivity and their breeding history. By avoiding inbreeding within the species, and cross-breeding with other species of tiger, the genetic integrity of the species can be maintained.

A combination of habitat protection and captive breeding is helping to conserve this important species.

EXAM PRACTICE

9 Outline the arguments for preserving biodiversity. [6]

10 a Name one intergovernmental organisation and one non-governmental organisation involved in conservation. [2]

   b Compare the roles and activities of these two organisations. [4]

11 Outline the general principles behind the World Conservation Strategy. [4]

12 Explain how the shape and size of a protected area can influence its success in protecting the organisms and ecosystems within it. [4]

13 Describe and evaluate captive breeding and reintroduction programmes as part of a species-based approach to conservation. [5]

14 Evaluate the strengths and weaknesses of the Convention on International Trade in Endangered Species (CITES). [4]

15 Name a protected area that you have studied and suggest five reasons that might explain why the area was selected for protection. [5]

Expert tip

You need to be able to evaluate different approaches to protecting biodiversity.

Water, aquatic food production systems and societies

4.1 Introduction to water systems

> **SIGNIFICANT IDEAS**
> - The hydrological cycle is a system of water flows and storages that may be disrupted by human activity.
> - The ocean circulatory system (ocean conveyor belt) influences the climate and global distribution of water (matter and energy).

The hydrological cycle

The world's water cycle is driven by the Sun. This allows the constant recycling of water between the oceans, the atmosphere and the land. This movement of water between these three reservoirs (storages) is known as the **'hydrological cycle'** or water cycle (Figure 4.1).

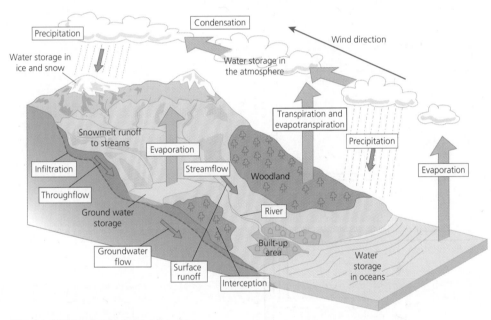

Figure 4.1 The hydrological cycle

▦ Freshwater and the Earth's water

▦ The oceans play a vital role in the hydrological cycle: 74% of the total precipitation falls over the oceans and 84% of the total evaporation comes from the oceans.

▦ Only a small fraction (2.5% by volume) of the Earth's water supply is freshwater Figure 4.2.

■ Storages of water

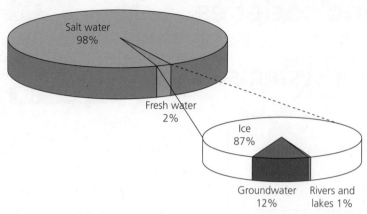

Figure 4.2 Freshwater sources

- Of the main stores of freshwater, over 87% is in the form of ice caps and glaciers, 12% is **groundwater** and the rest is made up of lakes, soil water, atmospheric water vapour, rivers and biota, in decreasing order of storage size.
- At a local scale the cycle has two main inputs – solar energy and precipitation (PPT) – and two major losses (outputs) – evapotranspiration (EVT) and runoff (Figure 4.3). A third output, leakage, can also occur from the deeper subsurface to other basins.

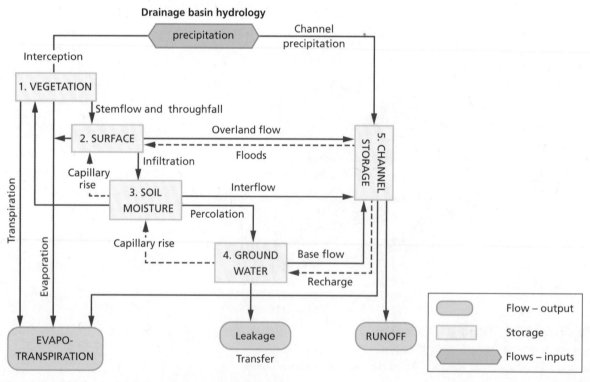

Figure 4.3 The hydrological cycle as a systems model

■ Flows in the hydrological cycle

Flows in the hydrological cycle include evapotranspiration (the combined loss of water through evaporation and transpiration), sublimation (ice changing to water vapour), condensation (water vapour changing to a liquid), evaporation (a liquid changing to a gas – water vapour), advection (wind-blown movement of water), precipitation (the movement of water from atmosphere to the land and oceans), melting (the conversion of solid snow and ice to liquid water), freezing (the change from a liquid form to a solid form such as ice).

Water on the surface may cause flooding (too much water to drain away), surface runoff (movement of water over the land's surface), infiltration (the movement of water from the surface downwards into the soil), percolation (the movement of water downwards from the soil to the rocks), and stream flow or currents (movement of water in channels such as streams and rivers).

■ Human impacts

Human activities such as agriculture, deforestation and urbanisation have a significant impact on surface runoff and infiltration. In general, human activities lead to a decline in vegetation cover, leading to reduced interception, more ground compaction and therefore more overland runoff. This increases the likelihood of floods, including flash floods. With urbanisation, permeable surfaces are replaced with impermeable surfaces, increasing the risk of flooding.

Ocean circulation systems

Ocean circulation systems are driven by differences in temperature and salinity. The resulting differences in water density drive the **ocean conveyor belt**, which transports heat and energy around the world, and thus affects climate.

> **Common mistake**
>
> Not all freshwater is accessible or renewable. Only freshwater lakes and rivers are renewable. Many groundwater reserves, such as those under the Sahara, are essentially non-renewable.

> **■ QUICK CHECK QUESTION**
>
> 1 What proportion of the Earth's freshwater is found in:
>
>   a groundwater
>
>   b lakes and rivers?

Revised ☐

> **Keyword definition**
>
> **Ocean conveyor belt** – The deep, large-scale circulation of the ocean's waters that is largely responsible for the transfer of heat from the tropics to colder regions.

Revised ☐

4.2 Access to freshwater

> **SIGNIFICANT IDEAS**
>
> • The supplies of freshwater resources are inequitably available and unevenly distributed, which can lead to conflict and concerns over water security.
>
> • Freshwater resources can be sustainably managed using a variety of different approaches.

Access to freshwater

Revised ☐

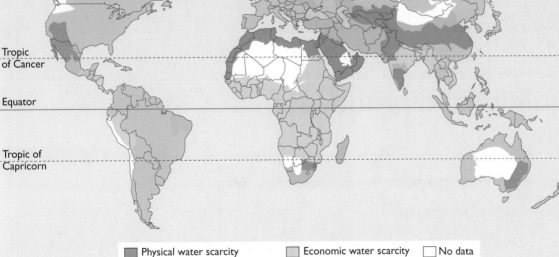

Figure 4.4 Areas of physical and economic water scarcity

Access to adequate amounts of safe freshwater varies considerably. Some areas have sufficient year-round rainfall (e.g. tropical rainforests), whereas others have water shortages that may be seasonal (e.g. areas with a monsoon climate) or year round (e.g. deserts).

■ The effect of climate change

It is likely that over the next few decades some areas will become drier as a result of global warming, and so water resources will be reduced.

■ Increasing demand for water

The concept of **water scarcity** relates water availability to the demands for water.

The demand for water increases with population growth, the need for more food, industrialisation and urbanisation. Globally, about 70% of water use is for farming and about 20% for industry. As societies become richer, the demand for water increases.

■ Contamination and over-extraction of water

Over-use of freshwater can lead to a decline in quantity (e.g. falling groundwater levels) and a decline in quality (e.g. eutrophication, salinisation). Hence, some water resources are not renewable.

■ Increasing supplies and conservation of water

There are many options for increasing water supplies. These include some large-scale, expensive measures such as the construction of large dams, water transfer schemes, desalination and the recharge of aquifers. There are other, less expensive measures such as the use of water butts, recycling of water and the use of 'grey water' for activities such as car-washing.

■ Water resources and conflict

Water resources have the potential to lead to international conflict. Conflicts may arise over the use of water in upper basin countries for hydopower or for irrigation, and the reduced availability of water resources in lower basin countries. Such conflicts include Egypt and Ethiopia over the Nile, India and Bangladesh over the Ganges, and China, Laos, Cambodia and Vietnam over the Mekong.

EXAM PRACTICE

1 Figure 4.4 shows the global distribution of physical and economic water scarcity (shortage).
   a Define the term *water scarcity*. [1]
   b Describe the pattern of global water scarcity. [3]
2 Evaluate the relative importance of factors that determine the sustainable use of freshwater resources. Refer to at least one case study in your answer. [9]

Keyword definitions

Physical water scarcity – Where water resource development is approaching, or has exceeded, unsustainable levels. It relates water availability to water demand and implies that arid areas are not necessarily water scarce.

Economic water scarcity – Where water is available locally, but not accessible for human, institutional or financial capital reasons.

Common mistake

There is no water scarcity in deserts. The concept of water scarcity relates water use to human demand. Because of the relative lack of people in deserts (because there is little water available) there is little demand for water, and therefore no water scarcity.

Expert tip

You must be able to **evaluate** the strategies that can be used to meet an increasing demand for freshwater.

You must be able to **discuss**, with reference to a case study, how shared freshwater resources have given rise to international conflict.

■ QUICK CHECK QUESTION

2 Suggest *two* reasons why freshwater supplies might be insufficient to meet the demands of human societies in the future.

4.3 Aquatic food production systems

Revised ☐

SIGNIFICANT IDEAS
• Aquatic systems provide a source of food production.
• Unsustainable use of aquatic ecosystems can lead to environmental degradation and collapse of wild fisheries.
• Aquaculture provides potential for increased food production.

Demand for aquatic food resources

The demand for aquatic food resources continues to increase as the human population grows and as standards of living rise, and changes in diet occur. Some of these changes are due to concerns about 'healthy eating' and some are due to changes in fashion.

■ Photosynthesis and aquatic food webs

Photosynthesis by phytoplankton supports a wide range of food webs, including those in the open ocean, those in continental shelf areas and those in shallow coastal seas.

The highest rates of productivity are found in shallow seas and near coastal areas, where the upwelling of nutrients enriches the surface waters, for example off the coast of Peru.

Terrestrial and aquatic food production systems (Figure 4.5) vary in terms of their trophic levels and efficiency of energy conversion.

- In terrestrial systems, most food is harvested from relatively low trophic levels (producers and herbivores).
- In aquatic systems most food is harvested from higher trophic levels where the total storages are much smaller. Although energy conversions along the food chain can be more efficient in aquatic systems, the initial fixing of available solar energy by primary producers tends to be less efficient due to the absorption and reflection of light by water.

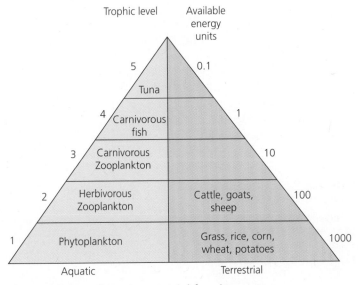

Figure 4.5 Aquatic and terrestrial farming systems

■ Biorights and selected species

Wild fisheries are referred to as 'capture fisheries', whereas aquaculture is the rearing and harvesting of fish in coastal and inland areas. However, the harvesting of certain species, such as whales, dolphins and bluefin tuna, can be controversial for different reasons, including biorights and the manner in which the animals are killed.

- The **maximum sustainable yield** is the largest amount of fish that can be captured without reducing the annual replacement of biomass. However, this is difficult to determine.
- Traditionally, indigenous peoples, such as the Inuit, hunted whales and seals within sustainable limits. There has been much controversy regarding Japan's decision to continue whaling for 'scientific purposes'.

Keyword definition

Maximum sustainable yield – The rate of increase in natural capital, i.e. that which can be exploited without depleting the original stock or its potential from replenishment.

■ Developments in the modern fishing industry

Developments in the fishing industry have led to overfishing in many areas, such as the Grand Banks off Newfoundland and the North Sea off the UK. It has been described as 'too many fishermen chasing too few fish'. Modern, industrialised intensive fishing uses satellites and sonar to track fish, long nets and trawler bags, and on-board refrigeration. Some trawlers can stay at sea for up to 3 months, and may fish in increasingly distant seas such as off the coast of West Africa or the Southern Ocean. The result has been falling fish stocks in many of the world's traditional fishing areas. Some small-scale practices can be very damaging, such as the use of dynamite on coral reefs to scare the fish into the open water.

Unsustainable and sustainable fishing practices

Revised ▢

It is possible to reduce some of the impacts of modern fishing. International policies, such as the European Union's Common Fisheries Policy, set out regulations for the size of catch permitted, the size of fish allowed to be caught, the size of mesh and the size of nets. National governments may establish marine parks and protected areas, for example the Soufrière Marine Management Area, St Lucia. Consumers can make sure that they buy fish that has been certified by the MSC (Marine Stewardship Council).

■ Aquaculture

The contribution of aquaculture to total fish production has increased dramatically in recent decades. It is likely that as yields from capture fisheries continue to fall, the contribution from aquaculture will increase. China is the world's largest aquaculture producer.

■ Issues surrounding aquaculture

Nevertheless, there are environmental issues connected with the growth of aquaculture including the loss of habitat, pollution caused by the use of fish feed, antibiotics and the spread of diseases within the farmed fish, but also into the wild population.

Common mistake

The focus of productivity and/or efficiency may be over simplistic. For example, the increased production of prawns from aquaculture is excellent for feeding people, but it has led to the destruction of about one-third of the world's mangrove forests, and this can have a knock-on effect on fishery production in nearby ecosystems.

Expert tip

You should be able to **discuss**, with reference to a case study, the controversial harvesting of a named species.

You should be able to **evaluate** strategies that can be used to avoid unsustainable fishing.

■ QUICK CHECK QUESTION

3 Which are more efficient, terrestrial ecosystems or aquatic ecosystems? Give reasons to support your answer.

4.4 Water pollution

Revised ▢

SIGNIFICANT IDEAS

• Water pollution, both to groundwater and surface water, is a major global problem the effects of which influence human and other biological systems.

Sources of freshwater and marine pollution

Revised ▢

There are a variety of freshwater and marine pollution sources. Sources of freshwater pollution include runoff, sewage, industrial discharge and solid domestic waste. Sources of marine pollution include rivers, pipelines, the atmosphere and activities at sea (operational and accidental discharges).

■ Types of aquatic pollutant

There is also a wide range of aquatic pollutants including floating debris, organic material, inorganic plant nutrients (nitrates and phosphates), toxic metals, synthetic compounds, suspended solids, hot water, oil, radioactive pollution, pathogens, light, noise and biological pollutants (**invasive species**).

■ Methods of testing water quality

Water quality can be measured using standard water testing kits. These kits include tests for dissolved oxygen, pH, phosphates, nitrates, chlorides and ammonia. The readings can then be compared with standardised charts, colour charts or tables of known samples. For example, when testing water for nitrates, clean water has less than $5 \, mg \, dm^{-3}$, whereas polluted water can contain $5–15 \, mg \, dm^{-3}$ (Table 4.1). It is important to compare different sites, for example upstream and downstream of a sewage outlet or factory.

Table 4.1 Typical values of selected indicators for clean water and polluted water

Substance	Clean water	Polluted water
Dissolved oxygen	Healthy, clean water generally has >75% oxygen saturation	Polluted water shows 10–50% oxygen saturation (raw sewage less than 10%)
Phosphate	Clean water contains $<5 \, mg \, dm^{-3}$	Polluted water contains $15–20 \, mg \, dm^{-3}$
Nitrate	Clean water contains $4–5 \, mg \, dm^{-3}$	Polluted water contains $5–15 \, mg \, dm^{-3}$
Ammonia	Clean water contains $0.05–0.99 \, mg \, dm^{-3}$	Polluted water contains $1–10 \, mg \, dm^{-3}$ (raw sewage contains $40 \, mg \, dm^{-3}$)

Common mistake

Water can contain ammonia, nitrates and phosphates and not be polluted – as long as it does not have too much of any of these.

■ QUICK CHECK QUESTION

4 Compare the dissolved oxygen content of clean water with that of polluted water.

Biodegradation of organic material

Revised ▢

Biodegradation of organic material utilises oxygen, which can lead to anoxic conditions and subsequent anaerobic decomposition, which in turn leads to formation of methane, hydrogen sulfide and ammonia (toxic gases).

■ Biochemical oxygen demand (BOD)

Aerobic organisms use oxygen in respiration. When there are more organisms and faster respiration, more oxygen will be used. Thus the **biochemical oxygen demand** at any point is affected by:

- the number of aerobic organisms
- their rate of respiration.

BOD is an indirect measurement used to assess pollution levels in water. The presence of an organic pollutant, such as sewage, causes an increase in the population of organisms that feed on and break down the pollutant. Organic pollution causes a high BOD. Certain species, such as bloodworms and *Tubifex* worms, are tolerant of organic pollution and the low oxygen content associated with it. In contrast, mayfly nymphs and stonefly larvae are associated with clean water (Figure 4.6).

Keyword definition

Biochemical oxygen demand (BOD) – A measure of the amount of dissolved oxygen required to break down the organic material in a given volume of water through aerobic biological activity.

Thermal pollution lowers the dissolved oxygen content of water. This leads to less aerobic respiration, increased anaerobic respiration and increased decomposition.

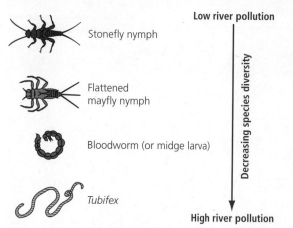

Figure 4.6 Invertebrate indicators of freshwater pollution

The effects of organic pollution are summarised in Figure 4.7.

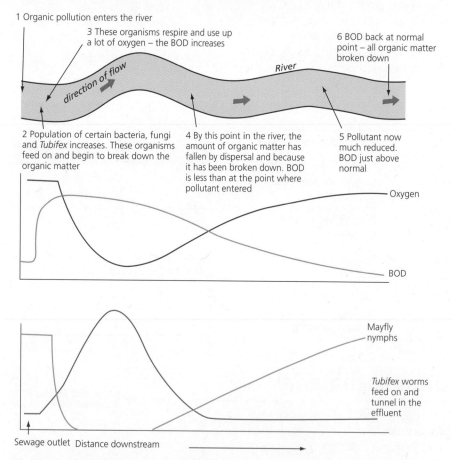

Figure 4.7 Effects of organic pollution

Common mistake

Sewage effluents do not necessarily discharge polluted water. Sometimes the water is treated and clean. If the water is polluted, the level of pollution will depend upon water temperature and how much water is in the stream.

Expert tip

In highly polluted rivers and streams, species diversity is low, although the population of a certain species may be high.

■ QUICK CHECK QUESTIONS

5 Define the term *biochemical oxygen demand* (BOD).

6 Describe the changes that occur downstream of a sewage outlet pipe.

7 Outline the effect of thermal pollution on dissolved oxygen.

Indicator species

Revised ☐

Certain species have different levels of tolerance to environmental conditions and change. The presence or absence and health of these indicator species can be used to suggest conditions in the environment.

Indicator species can be used as an indirect measure of pollution and/or environmental degradation. Examples include lichens, nettles, red algae and freshwater invertebrates such as *Tubifex* worms, bloodworms and mayfly nymphs.

- Certain lichen species (such as *Usnea articulata*) indicate very low levels of sulfur dioxide in the atmosphere.
- Nettles (*Urtica dioica*) indicate high phosphate levels in the soil.
- Red alga (*Corallina officinalis*) inhabits saline rock pools and is absent from brackish pools.

Biotic indices

Revised ☐

Biotic indices can be used as an indirect method of measuring pollution levels. This involves levels of tolerance, diversity and abundance of organisms.

■ Trent biotic index

The Trent biotic index is based on the disappearance of certain indicator species as the levels of pollution increase (Table 4.2). Changes in the amount of light and dissolved oxygen cause less-tolerant species to die out. As pollution increases, diversity decreases. The maximum value for the Trent biotic index is 10.

Table 4.2 Indicator species for the Trent biotic index

Indicator species	Number present	Total number of groups present				
		0–1	2–5	6–10	11–15	16+
		Trent biotic index				
Stone fly (*Plecoptera*) nymph present	More than one species	–	7	8	9	10
	One species only	–	6	7	8	9
Mayfly (*Ephemeroptera*) nymph present	More than one species	–	6	7	8	9
	One species only	–	5	6	7	8
Caddis fly (*Trichoptera*) larvae present	More than one species	–	5	6	7	8
	One species only	4	4	5	6	7
Gammarus present	All above species absent	3	4	5	6	7
Shrimps – crustacean (*Aseilus*) present	All above species absent	2	3	4	5	6
Tubifex worms and/or red bloodworm (*chironomid*) larvae present	All above species absent	1	2	3	4	–
All above types absent	Some organisms (e.g. *Eristalis tenax*) not requiring dissolved oxygen may be present	0	1	2	–	–

Expert tip

When monitoring a stretch of water, try to examine seasonal changes as well as changes above and below the site.

■ **QUICK CHECK QUESTIONS**

8 What is the minimum Trent biotic index for a sample of water containing stonefly nymphs?

9 What is the maximum Trent biotic index for a water sample containing *Tubifex*?

EXAM PRACTICE

3 Define the term biochemical oxygen demand (BOD) and explain how this indirect method is used to assess pollution levels in water. [5]

Eutrophication

Revised ▢

■ The process of eutrophication

Natural and human causes of eutrophication:

■ There are natural and human causes of **eutrophication**.

■ Both result in an increase in nitrates and/or phosphates, leading to rapid growth of algae, accumulation of dead organic matter, a high rate of decomposition and lack of oxygen.

■ Natural cycles can include nutrients added from decomposing biomass, runoff from surrounding areas and upwelling ocean currents bringing nutrients to the surface.

■ Human causes include runoff of fertilisers or manure from agricultural land. Similarly, domestic wastewater may contain phosphates from detergents, and non-treated sewage can also lead to eutrophication.

Figure 4.8 A eutrophic river

■ **QUICK·CHECK QUESTIONS**

10 Identify *two* natural causes of eutrophication.

11 Identify the *two* main chemicals involved in the process of eutrophication.

Common mistake

Eutrophication can be natural as well as human-induced. Human-induced eutrophication happens a lot faster than natural eutrophication and occurs on a larger scale.

> **Keyword definition**
>
> **Eutrophication** – The natural or artificial enrichment of a body of water, particularly with respect to nitrates and phosphates, that results in depletion of oxygen levels in the water. Eutrophication is accelerated by human activities that add detergents, sewage or agricultural fertilisers to bodies of water.

■ Eutrophication and feedback cycles

Table 4.3 Positive and negative feedback in the eutrophication process

Positive feedback occurs in the process of eutrophication	Negative feedback can also occur during the process of eutrophication
● As more nutrients are added to the system, the biomass of algae increases due to the availability of nutrients.	● The increase in nutrients promotes growth of plants that store them in biomass.
● Decomposition of the increased biomass leads to further nutrient load and so further deviation from the long-term equilibrium.	● This leads to a reduction in nutrients, so balance is restored, i.e. negative feedback.
● Hence positive feedback occurs.	● The increase in algae will lead to increases in species that feed on algae.
● The growth of algae blocks light, so causing underwater plants to die and create more nutrients as they decompose.	● This may lead to subsequent decrease in algal populations, so balance is restored.
● More nutrients lead to further growth of algae so further deviation from the equilibrium, thus there is increased positive feedback.	● The increase in dead organic matter provides more food for decomposers, which increase in number.
● An increase in bacteria causes increased BOD, hence oxygen-dependent organisms to die.	● Increased rate of decomposition leads to a decrease in dead organic matter, so balance is restored, i.e. negative feedback occurs.
● This increase in dissolved organic matter leads to even further increase in BOD/bacteria, so further deviation from equilibrium, reinforcing positive feedback.	

■ The impacts of eutrophication

Impacts on the ecosystem:

- Both human and natural eutrophication lead to an increase in biomass of algae.
- The algae lower light penetration to underwater plants.
- The increased death of algae and underwater plants leads to an increase in bacteria and a reduction in oxygen levels, leading in turn to a decrease in biodiversity.
- More bacteria increase BOD, which leads to lowered oxygen content of water (hypoxia).
- Reduced oxygen leads to the death of many organisms.
- Net primary productivity is usually higher compared with unpolluted water, and may be indicated by extensive algal or bacterial blooms.
- Diversity of primary producers changes and finally decreases; the dominant species change.
- The length of the food chain decreases as algae lock up the nutrients and block sunlight from reaching the river or lake bed.
- However, as eutrophication proceeds, early algal blooms give way to cyanobacteria (blue-green algae).
- Fish populations are adversely affected by reduced oxygen availability, and the fish community becomes dominated by surface-dwelling coarse fish, such as roach and rudd.
- With less sunlight penetrating the water macrophytes (submerged aquatic plants) disappear because they are unable to photosynthesise.
- In theory, the submerged macrophytes could also benefit from increased nutrient availability, but they are shaded by the free-floating microscopic organisms.

■ Impacts on society

Eutrophication has impacts on human populations. One is financial – the loss of fertilisers from fields may reduce crop productivity and therefore farm yield and profit. Moreover, the cost of treating nitrate-enriched water is expensive – in the UK this is estimated to cost between £50 million and £300 million each year.

Nitrate-enriched water is associated with higher rates of stomach cancer and 'blue baby syndrome' (methaemoglobanaemia – insufficient oxygen in pregnant women's blood). However, correlation does not equate with causation and many other factors are likely to be involved.

Expert tip

For full credit in describing feedback, responses should identify two steps:

- Step 1: how change in one factor causes change in another.
- Step 2: how the second factor affects the first, either increasing its change (positive feedback) or decreasing its change (negative feedback).

■ QUICK CHECK QUESTIONS

12 Explain why macrophytes die out as a result of eutrophication.

13 Explain why BOD increases following eutrophication.

Common mistake

Eutrophication leads to a change in species composition rather than removing all species. Surface-dwelling organisms are favoured rather than bottom-dwelling organisms, with the exception of the bacteria that decompose the dead organic matter.

Expert tip

Most lakes are naturally oligotrophic (nutrient poor) – once eutrophication starts to occur, productivity increases as nutrient enrichment occurs.

■ Dead zones

Dead zones can occur in oceans and freshwater when there is not enough oxygen to support aquatic and marine life. They can be the result of eutrophication.

Pollution management strategies

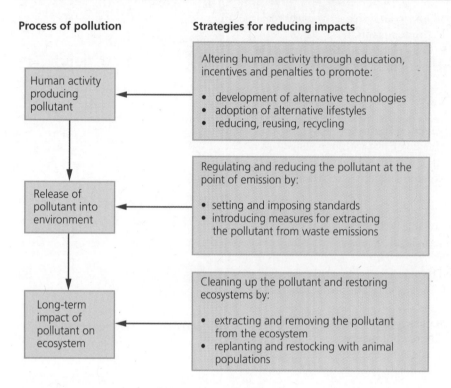

Figure 4.9 Approaches to pollution management

There are many pollution management strategies (Figure 4.9), some of which are outlined below.

■ Altering human activity

Altering the human activity that produces pollution can include alternative methods of enhancing crop growth, alternative detergents, and so on. For example:

- Avoid spreading fertilisers in winter, as the soil is bare and runoff may wash fertilisers in to rivers and streams.
- Use autumn-sown crops – these maintain a cover during the winter, and may conserve nitrogen in the soil.
- Use less nitrogen if the previous year was dry (more will be left in the soil).
- Use organic fertilisers (manure) on agricultural fields.
- Practise mixed cropping or crop rotation so that less fertiliser is needed.
- Do not apply fertilisers to fields that are next to streams and rivers.

■ Regulating and reducing pollutants at the point of emission

Ways to regulate and reduce pollutants at the point of emission include the following:

- Use sewage treatment processes that remove nitrates and phosphates from the waste.
- Use zero- or low-phosphorus detergents.
- Only use washing machines with a full load of washing.
- Reduce the use of fertilisers in gardens.

■ Clean-up and restoration

Clean-up and restoration techniques include the following:

- Pump nutrient-rich sediment from eutrophic lakes and reintroduce plant and fish species.
- Pump air through lakes to avoid the low-oxygen conditions.

■ QUICK CHECK QUESTIONS

14 Explain why fertilisers should not be used in winter.

15 What type of pollution management strategy is phosphate stripping?

- Remove biomass (e.g. water hyacinth or reed) and use it for biofuel or for thatching.
- Reintroduce plant and animal species.
- Use barley bales to lock up nitrates in the water (Figure 4.10).
- Treat with a solution of aluminium or ferrous salt to precipitate phosphates.

Expert tip

In some cases the same response can be classified as technocentric or ecocentric. Barley bales are an ecocentric approach (using nature), although the research into the best forms of material to absorb nitrates can be a technocentric approach.

Common mistake

Although there are many potential solutions to the problem of eutrophication, it is not always possible to solve it. This is because it is non-point-source pollution and there are many different causes. Moreover, not every country has the resources to implement the strategies.

Figure 4.10 The use of barley bales to lock up nitrates in the water

EXAM PRACTICE

4   Describe and evaluate ecocentric and technocentric responses to eutrophication.   [8]

Expert tip

Using Figure 4.9, you should be able to show the value and limitations of each of the three different levels of intervention. In addition, you should appreciate the advantages of employing the earlier strategies over the later ones and the importance of collaboration in the effective management of pollution.

Expert tip

You should be able to:

- **analyse** water pollution data
- **explain** the process and impacts of eutrophication
- **evaluate** the uses of indicator species and biotic indices in measuring aquatic pollution
- **evaluate** pollution management strategies with respect to water pollution.

Common mistake

Some pollution can be natural – volcanic eruptions can cause acidification and climate change – it is not entirely due to humans.

Soil systems and terrestrial food production systems and societies

5.1 Introduction to soil systems

> **SIGNIFICANT IDEAS**
> - The soil system is a dynamic ecosystem that has inputs, outputs, storages and flows.
> - The quality of soil influences the primary productivity of an area.

The soil system

Soil is the outermost layer of the Earth's surface, consisting of weathered bedrock (**regolith**), air, water and both living and dead organic matter. Soil systems link the soil with the lithosphere, atmosphere and living organisms. Figure 5.1 illustrates some of these links.

> **Keyword definition**
> **Soil** – A mixture of mineral particles and organic material that covers the land, and in which terrestrial plants grow.

▆ Storages

Soil system storages include organic matter, organisms, nutrients, minerals, air and water.

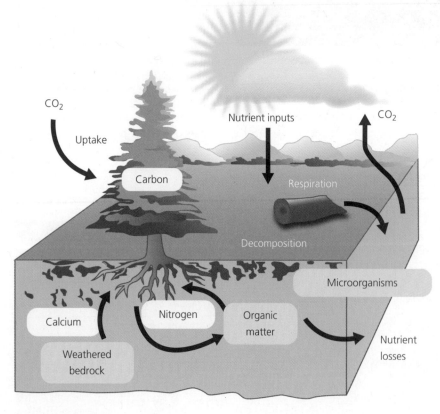

Figure 5.1 Soil systems, lithosphere systems and atmospheric systems

▆ Transfers

Soil processes are directly affected by atmospheric process. The amounts of heat, evaporation and precipitation determine the main movements of water within the soil.

Lithospheric systems are also important – the **parent material** (bedrock) influences soil drainage and **soil fertility**.

The **soil profile** (Figure 5.2) is influenced by biotic factors. Earthworms help mix the soil, while fungi and bacterial activity help break down plant litter and form **humus**.

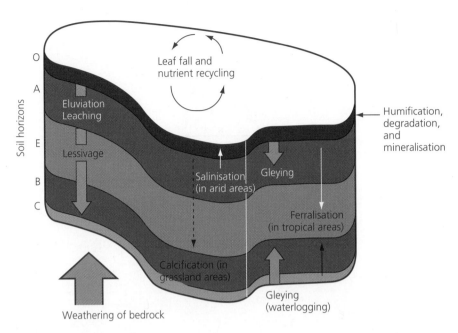

Figure 5.2 A soil profile

■ The **soil profile** shows distinct **soil horizons** – these include O (organic horizon), A (mixed mineral/organic horizon), E (eluvial or leached horizon), B (illuvial or deposited horizon) and C (bedrock or parent material).
■ Transfers of material (including deposition) result in reorganisation of the soil.
■ **Leaching** generally refers to the removal of material down a soil through solution and suspension. (Leaching, **illuviation** and **eluviation** are often used interchangeably but they have precise definitions.)

Inputs and outputs

■ There are inputs of organic and parent material, precipitation, infiltration and energy.
■ Outputs include leaching, uptake by plants and mass movement.

Transformations

■ Transformations include decomposition, weathering and nutrient cycling.
■ **Humification, degradation** and **mineralisation** form the process whereby organic matter is broken down and the nutrients are returned to the soil. The breakdown releases organic acids – chelating agents – which break down clay to silica and soluble iron and aluminium.
■ **Illuviation** is the redeposition of material in the lower horizons.

> ### Keyword definition
> **Soil profile** – A vertical section through a soil, from the surface down to the parent material, revealing the soil layers or horizons.

■ QUICK CHECK QUESTIONS

1 Define the term *soil*.
2 Identify *two* ways in which bedrock influences soils, and *two* ways in which climate affects soils.

Soil structure and texture

Revised ▢

■ **Soil structure** refers to the shape of the particles. It has an effect on primary productivity.
■ **Soil texture** refers to the size of soil particle.
■ Sand particles are those less than 2 mm in diameter, silt particles are less than 0.02 mm and clay less than 0.002 mm.

■ Soil texture triangular graph

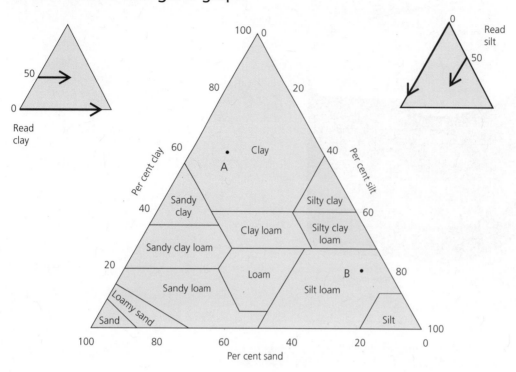

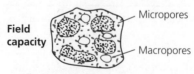

Figure 5.3 A triangular graph to show soil textural groups

■ Soil textural groups are often shown by the use of triangular graphs (Figure 5.3).

■ Soil texture is important as it affects: moisture content and aeration; retention of nutrients; and ease of cultivation and root penetration.
■ Clay soils have a large surface area in relation to volume and so have a high potential for exchange of nutrients, but they become waterlogged, and are described as 'cold' or 'heavy'. In years of drought the shrinkage of some clay soils can cause structural damage.
■ Sandy soils drain rapidly, and are described as 'light'.
■ Silt soils are especially prone to compaction if ploughed when wet.
■ The maximum amount of water that a soil can hold is referred to as **field capacity** (Figure 5.4).

Expert tip

Make sure that you can draw diagrams such as Figures 5.1 and 5.2 to show the soil system and how processes might take place within a soil.

Common mistake

Triangular graphs are often read incorrectly – the sum of the percentages must add up to 100%.

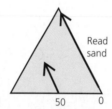

Saturation

- all pore spaces filled with water
- some water drains as a result of gravity

Field capacity

Micropores

Macropores

- small micropores filled with water and held by suction
- macropores (large pores) filled with air
- water available to plants

Wilting

- water present only in small quantities, held by soil hygroscopically

Figure 5.4 Soil moisture

A loam soil is often considered to be the best soil for cultivation as it has the optimum combination of sand, silt and clay. It is therefore easily workable, drains well, retains moisture and nutrients and is well aerated. As a result, it has the highest plant productivity.

Table 5.1 Storage capacity of soils (cm water per 30 cm depth)

Soil texture	Field capacity	Wilting point	Available water
Sandy loam	5.6	2.8	2.8
Loam	8.4	4.3	4.1
Clay loam	9.9	5.3	4.6
Heavy clay	11.9	6.3	5.6

■ QUICK CHECK QUESTIONS

3 Identify the percentage of sand, silt and clay found in soils A and B in Figure 5.3.
4 State what textural group each of the following soils belongs to:
   a 60% sand, 30% clay and 10% silt
   b 40% sand, 40% silt and 20% clay
5 Which of the soils in Table 5.1 is more likely to require irrigation to help plant productivity?

Expert tip

You should be able to:

- **outline** the transfers, transformations, inputs, outputs, flows and storages within soil systems
- **explain** how soil can be viewed as an ecosystem
- **compare and contrast** the structure and properties of sand, clay and loam soils, with reference to a soil texture diagram, including their effect on primary productivity.

5.2 Terrestrial food production systems and food choices

Revised ■

SIGNIFICANT IDEAS
- The sustainability of terrestrial food production systems is influenced by socio-political, economic and ecological factors.
- Consumers have a role to play through their support of different terrestrial food production systems.
- The supply of food is inequitably available and land suitable for food production is unevenly distributed among societies, and this can lead to conflict and concerns.

Sustainability

Revised ■

The sustainability of terrestrial food production systems is influenced by many factors including:

■ scale
■ industrialisation/mechanisation
■ use of fossil fuel for transport, fertilisers and for machinery
■ use of genetically modified organisms versus organic farming
■ water use
■ fertilisers
■ pest control
■ pollinators
■ antibiotics
■ government policy
■ commercial versus subsistence food production.

Inequalities

Revised

There are many differences in food production and distribution around the world, which are affected by a range of socio-political, economic and ecological influences.

- The world is growing by over 80 million people each year, hence more food is needed.
- In addition, a larger proportion of people in the world will become middle- and high-income earners, and there will be a corresponding change in diet from grain-based to meat- and dairy-based.

During the latter part of the twentieth century, the growth in food production out-paced the growth of population. This was largely due to the **Green Revolution** (the application of science and technology to agriculture, leading to high-yielding varieties, breeding programmes, widespread use of chemical fertilisers and pesticides, irrigation and so on).

However, between 2001 and 2008 world food prices rose dramatically. Although there was an increase in food production, there was an even greater increase in demand for food. A rise in oil prices led to higher costs in transport and fertiliser production.

Food distribution

Revised

Once food is produced it needs to be brought to the market to be sold. Sometimes food is collected from the farmer whereas at other times the farmer has to get it to the market. This transportation uses fossil fuels. The distance that a food travels to its destination is known as **food miles**. Some foods are transported huge distances.

In addition, all of the inputs, such as fertilisers, machinery and equipment, need to be transported to the farmer. These are expensive and also use up fossil fuels.

Expert tip

You must have an example of *one* commercial farming system and *one* subsistence system.

Common mistake

It is not just population growth that causes an increased demand for food – standard of living is important too. The gradual change in diet by people in newly rich and industrialising countries (such as Korea and, increasingly, China and India) is one of the most important factors leading to a relative food shortage.

■ QUICK CHECK QUESTIONS

6 How was agricultural production in the twentieth century able to grow more rapidly than population growth?

7 Suggest reasons for the falling level of productivity after about 2004.

■ Food waste

Food waste is prevalent in both LEDCs and more economically developed countries (MEDCs), but for different reasons. In LEDCs most food waste occurs either at the farm level, due to lack of storage or refrigeration, or in transport to the market. In contrast, most food waste in MEDCs occurs at a household level or at a retail level (e.g. 'sell-by' and 'best-before' dates).

Factors influencing food production systems

Revised

There are many factors that influence food production systems. These include socio-economic, cultural, ecological, political and economic factors:

- Socio-economic factors may include level of education and skills.
- Cultural factors may include religious impacts, for example most Hindus do not eat beef, and Islam and Judaism forbid the consumption of pork.
- Ecological factors may prevent a farming system from developing, for example dry areas favour nomadic pastoralism rather than permanent agriculture.

■ Political factors may include government subsidies to promote production of certain food types, or agricultural policies, which protect farmers by preventing imports of different food types.

■ Economic factors may influence the level of mechanisation that farmers can afford, but it also influences the demand for certain food types – diets change as people become more prosperous.

■ Land availability

The amount of land that is available for farming is decreasing in many places. This is partly due to population growth and urbanisation, using up productive land for human settlement. In addition, industrial development, mining and the building of dams and roads reduces the amount of land available for farming. Soil erosion and degradation reduce the quality of soil. This makes the amount of land available per person much lower now than in the past, and therefore has a potential impact on food production.

Another cause is the use of land for the production of biofuels rather than food crops. Approximately 100 million tonnes of grain are used for biofuels. As more grain is used for biofuel, less grain (and land) is used for the production of food for human use.

Food yield from different trophic layers

Revised ☐

The yield of food per unit area from lower trophic levels is greater in quantity, lower in cost and may require fewer resources – for example it is cheaper and easier to raise wheat and corn than livestock.

Terrestrial and aquatic food production systems (Figure 4.5 on page 95) vary in terms of their trophic levels and efficiency of energy conversion:

■ In terrestrial systems, most food is harvested from relatively low trophic levels (producers and herbivores).

■ In aquatic systems most food is harvested from higher trophic levels where the total storages are much smaller. Although energy conversions along the food chain can be more efficient in aquatic systems, the initial fixing of available solar energy by primary producers tends to be less efficient due to the absorption and reflection of light by water.

> **Expert tip**
>
> If a food has more food miles, this does not mean it is more harmful to the environment than one that is consumed locally. A locally grown food might use more fertiliser and irrigation water and provide less return per unit of input.

> **Common mistake**
>
> The focus on productivity and/or efficiency may be over-simplistic. For example, prawn farming has led to the destruction of about one-third of the world's mangrove forests and this can have a knock-on effect on fishery production in nearby ecosystems.

Cultural choices

Revised ☐

Many people choose to eat from higher up the food chain, i.e. meat products and dairy products, and this requires a greater amount of land and inputs. This is associated with increasing levels of wealth and rising standards of living. In many cases, increasing intake of meat and dairy products represents a higher standard of living than one based on staples such as maize or rice.

Terrestrial food production systems

Revised ☐

Terrestrial farming systems can be classified according to four pairs of contrasting categories:

■ **arable** (the cultivation of crops, e.g. the corn belt in the USA) or **pastoral** (the rearing of animals, e.g. the Maasai herdsmen of Kenya)

- **commercial** (products are sold to make a profit, e.g. market gardening in the Netherlands) or **subsistence/peasant** (products are consumed by the cultivators, e.g. shifting cultivation by the Kayapo of the Amazonian rainforest)
- **intensive** (high inputs or yields per unit area, e.g. cattle feed lots in California) or **extensive** (low inputs or yields per unit area, e.g. reindeer herding in Siberia)
- **nomadic** (farmers move seasonally with their herds, e.g. the Pokot pastoralists in Kenya) or **sedentary** (farmers remain in the same place throughout the year, e.g. rice farmers in Southeast Asia).

> **Expert tip**
>
> The classification of farming into four dichotomous ('either/or') categories is an over-simplification. A subsistence farmer may sell a small amount of produce whereas a commercial farmer may keep some farm produce for home consumption. Each pair should be seen as a sliding scale from one to the other.

■ QUICK CHECK QUESTIONS

8 Describe the dairy farming system and the nomadic pastoralist system using the four sets of contrasting terms described in the text.

9 Which are more efficient, terrestrial ecosystems or aquatic ecosystems? Give reasons to support your answer.

Contrasting food production systems

Revised ☐

Factors to be considered when contrasting two food production systems (Figure 5.5) should include:

- inputs – for example, fertilisers (artificial and organic), irrigation water, rainfall, pesticides (natural and artificial), fossil fuels, food distribution, human labour and mechanisation, seeds (GMO and traditional), breeding stock (domestic and wild), antibiotics and hormones
- outputs – including food quantity and quality, pollutants (nitrates, fertilsers), chemicals from pesticides and herbicides, emissions from machinery, seeds, animals for breeding
- system characteristics – for example, selective breeding, genetically engineered organisms, monoculture versus polyculture, sustainability, indigenous versus introduced crop species
- socio-cultural – for example, for the Maasai cattle equals wealth, and quantity is more important than quality; farming for subsistence or profit, for local consumption or export, for quality or quantity
- environmental impact – for example, pollution, habitat loss, reduction in biodiversity, soil erosion.

> **Expert tip**
>
> When contrasting two systems, those you choose should be both terrestrial or both aquatic. In addition, the inputs and outputs of the two systems should differ qualitatively and quantitatively (although not all systems will be different in all aspects). A pair of examples could be, for example: North American cereal farming and subsistence farming in some parts of Southeast Asia; intensive beef production in the developed world and the Maasai tribal use of livestock; or commercial salmon farming in Norway/Scotland and rice/fish farming in Thailand. Other local or global examples are equally valid.

Table 5.2 Factors affecting agriculture

Physical factors	Socio-economic factors
Precipitation: type, frequency, intensity, amount	Land tenure/ownership: ownership, rental, share-cropping, state-control
Temperature: growing season (>6°C), ground frozen (0°C), range of temperatures	Organisation: collective, cooperative, agribusiness, family farm
Soil: fertility (pH, cation exchange capacity), nutrient status, structure, texture, depth	Government policies: subsidies, guaranteed prices, ESAs (environmentally sensitive areas), quotas, set-aside
Pests: vermin, locusts, disease	War, disease, famine
Location: slope gradient, relief, altitude, aspect – ubac (shady) or adret (sunny)	Farm size: field size and shape
	Demand: size and type of market
	Capital: equipment, machinery, seed, money, 'inputs'
	Technology: HYVs, fertilisers, irrigation
	Infrastructure: roads, communications, storage
	Advertising
	Cultural and traditional influences
	Education and training
	Behavioural influences
	Chance

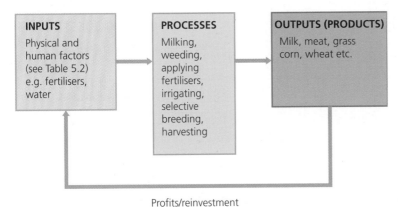

Profits/reinvestment

Figure 5.5 The farming system

CASE STUDY

SHIFTING CULTIVATION IN MEXICO

The Popoluca Indians of Santa Rosa, Mexico practise a form of shifting cultivation. This farming system is intensive, subsistence, largely arable and semi-nomadic (they change plots every few years). Most of the labour is done by hand, with men doing the cutting, burning and hunting and women doing most of the farming.

The plots are quite small, typically 3–4 ha. These are cleared for farming by cutting down and burning trees, which enriches the soil with nutrients. However, most trees and shrubs are not burned, but are harvested for their fruits and seeds. The system produces over 250 different types of crop (polyculture). Those cultivated include coffee, oregano, squash and yams. The only input that is bought in is maize seed; the rest is from natural forest that is allowed to seed itself.

The men hunt for game, fish and turtles. There is some trading of surplus game. Fruit and insects are also gathered from the forest.

The system uses natural fertiliser (animal manure) as well as the burning of trees to enrich the soil. No irrigation water or pesticides are used, so no fossil fuels are needed. Energy ratios (energy outputs/energy inputs) may be as high as 65.

There are some environmental impacts – for example, the cutting down and burning of vegetation is believed to have long-term effects on the quality of the soil (see Figure 5.10).

CASE STUDY

INTENSIVE COMMERCIAL PIG PRODUCTION IN DENMARK

Pig farming in Denmark, an MEDC, is commercial, pastoral, intensive and sedentary. The system, based on the selective breeding, rearing and selling of stall-fed pigs, is heavily mechanised, with extensive use of food concentrates, machinery for feeding and transport services. Inputs also include veterinary services, the use of hormones to increase productivity and antibiotics to reduce the spread of disease.

Farms are generally between 10 ha and 30 ha in size. The work structure is increasingly based on a farm manager and hired labour. Many farms will also produce cereals and keep a dairy herd – skimmed milk and whey are fed to the pigs. The system is still classified as a monoculture because it is not diverse.

Energy ratios (outputs/inputs) are only around 0.4.

The system is a commercial one – over nine million pigs are produced annually. Up to 75% of Danish bacon is exported and this accounts for 43% of Danish agricultural export.

The environmental impacts of this system are high:

- There is much use of fossil fuel in the distribution of pig products around the world and in the running of machines. Fossil fuels are also used in the production of feed concentrates.

- A large amount of manure is produced, the smell of which can impact on the local population of humans. It also produces methane as it decomposes.

- Fertilisers, pesticides and irrigation water are used to improve crop yields.

Common mistake

Some people believe that shifting cultivation is 'in harmony' with the natural environment. It can have a negative impact on soils. In addition, as areas of tropical rainforest become smaller, the pressures on the remaining areas become greater.

Expert tip

Rather than just thinking about differences, try to identify some similarities between the two systems – for example, both rely on physical factors such as climate and soil fertility, although these may impact in different ways.

■ **QUICK CHECK QUESTIONS**

10 In what ways is shifting cultivation intensive?

11 How does this differ from intensive commercial food production?

Links between social systems and food production

Revised

There are many examples of links between social systems and food production systems.

■ In some LEDCs low population densities are associated with shifting cultivation and nomadic pastoralism.

■ In MEDCs such as the USA and Australia, low population densities may be associated with highly mechanised commercial farming.

■ In areas of high population densities, such as in Southeast Asia, intensive farming has resulted.

■ Nomadic pastoralism

■ Traditional nomadic pastoralists, such as the Pokot of Kenya, live out their lives in relative harmony with the natural environment. For example, the energy flow in the nomadic system is similar to that of the savannah ecosystem.

■ Herdsmen use the savannah grasslands to feed their herds.

■ Cattle replace wild herbivores as the main herbivores.

■ There is limited killing of wild predators.

■ The nutrient cycling system can be altered.

■ True nomadic movement returns and distributes nutrients over a wide area although some concentrations of nutrients, in dung, can occur if herds remain in one place for a length of time, such as at a borehole.

■ Biological productivity in savannah grasslands is low and variable.

■ NPP varies from about $150\,\mathrm{g\,m^{-2}\,yr^{-1}}$ in drier areas, rising to $600\,\mathrm{g\,m^{-2}\,yr^{-1}}$ in wetter margins.

■ Secondary productivity is low – hence farmers use milk, milk products and blood rather than meat.

■ Their animals are their source of wealth, so they only kill the very old or very sick.

■ Environmental impacts might be unsustainable.

■ Over-exploitation of grass or over-concentration of herds removes vegetation, especially sweeter species, causing ponding of the surface, gulleying and desertification – the spread of desert conditions (Figure 5.9 on page 116).

■ In traditional pastoral societies desertification has been due to climatic deterioration. Now, however, economic, social and political reasons are increasingly to blame. These have led to larger herds, shorter nomadic routes and greater pressure around water sources such as boreholes.

■ Modern agribusiness (agro-ecosystems)

■ Unlike traditional subsistence farming, where the farmers do not own the land but have rights to use it, in modern agriculture there is generally some form of ownership (which may or may not then be rented to another farmer).

■ Modern agribusiness is associated with capitalism and the drive to make profits.

■ Increasingly farming is being driven by the demands of large retailers to provide high volumes of uniform-quality food to an increasingly urban population.

■ Food production and distribution have gone global.

- Farming has become increasingly intensive, large-scale and globalised in the drive for cheaper food.
- Advances in technology and communications have combined with falls in the costs of transport to transform the way in which food is sourced.
- The concentration of power in retailing and food processing has affected those at the other end of the scale, namely farmers in LEDCs and small farmers in MEDCs.
- Increasingly, modern farming methods are having a negative impact on the environment.
- The term **agro-industrialisation** refers to the large-scale, intensive, high-input, high-output, commercial nature of much of modern farming.

Improved yields and environmental impacts

- Since the 1960s, wheat yields have increased from 2.6 to 8 tonnes per hectare and barley yields from 2.6 to 5.8 tonnes, while each cow produces twice as much milk.
- Cleaning up the chemical pollution, repairing the habitats and coping with disease caused by industrial farming costs up to £2.3 billion a year.
- It now costs water companies $165–250 million to remove pesticides and nitrates from drinking water.
- Food processors usually want large quantities of uniform-quality produce or animals at specific times. This is ideally suited to intensive farming methods, which favour synthetic chemicals, land degradation and animal welfare problems. For example, apples receive an average of 16 pesticide sprays.
- The global food industry has a massive impact on transport (Figure 5.6).
- The food system has become almost completely dependent on crude oil, making food supplies vulnerable, inefficient and unsustainable.

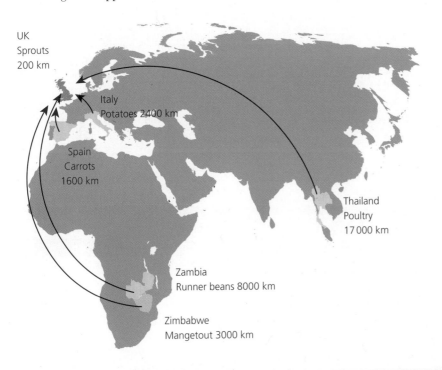

UK
Sprouts
200 km

Italy
Potatoes 2400 km

Spain
Carrots
1600 km

Thailand
Poultry
17 000 km

Zambia
Runner beans 8000 km

Zimbabwe
Mangetout 3000 km

Figure 5.6 The wastefulness of a Christmas dinner – the ingredients of a traditional Christmas meal bought from a supermarket may have cumulatively travelled 24 000 miles

CASE STUDY

WATER PROBLEMS AND GLOBAL FARMING IN KENYA

The shores of Lake Naivasha in the 'Happy Valley' area of Kenya have been seriously polluted. Environmentalists blame the problems on pollution from pesticides, excessive use of water on farms and deforestation caused by migrant workers in the growing shantytowns foraging for fuel.

British and European-owned flower companies in the area grow vast quantities of flowers and vegetables for export. Much water is used to produce flowers. The export of these flowers (a flower is 90% water) is effectively exporting Kenyan water. This is known as '**virtual water**'.

The greatest impact is being felt by the nomadic pastoralists in the semi-arid areas to the north and east of Mt Kenya. The flower farms have taken over land that the pastoralists used and there is now less water.

Food sustainability

Revised ▢

Increased food sustainability can be achieved in many ways:

- reduce the consumption of meat
- increase consumption of organically grown food
- increase consumption of locally produced goods
- improve food labelling to improve food choices
- monitor and control the standards and activities of multinational companies and food corporations
- introduce buffer zones around areas used for food production to absorb farm waste (e.g nutrient runoff).

5.3 Soil degradation and conservation

Revised ▢

Soil fertility and succession

Revised ▢

Soil ecosystems change through succession. Fertile soil contains a community of organisms that work to maintain functioning nutrient cycles and that are resistant to soil erosion. Fertile soil can be considered as a non-renewable resource because once depleted, it can take significant time to restore the fertility or in some cases it never recovers.

Soil degradation

Revised ☐

Human activities such as overgrazing, deforestation, urbanisation, unsustainable agriculture such as monoculture and irrigation cause processes of **soil degradation** (Figure 5.7). These processes include:

- **erosion** by wind and water. There are many types of water erosion including surface, gully, rill and runnel erosion.
- **biological degradation** (the loss of humus and plant/animal life)
- **physical degradation** (loss of structure and changes in permeability). Groundwater over-abstraction, for example, may lead to dry soils, leading to physical degradation.
- **chemical degradation** (acidification, declining fertility, changes in pH, salinisation and chemical toxicity). Acidification is a change in the chemical composition of the soil that can trigger the circulation of toxic metals.

Farming and soil fertility

The impact of farming on soil fertility varies. Large-scale, highly mechanised, commercial farming tends to have a greater impact on soil fertility than small-scale subsistence farming. For example, heavy machinery may compact soils, make them more impermeable and reduce the amount of oxygen that they contain. The use of fertilisers alters their chemical composition, and the use of pesticides and insecticides may lead to increased amounts of dangerous toxins in the soil. Subsistence farming can also damage soils – shifting cultivation may reduce soil fertility and it may take hundreds of years for the soil to regain its lost fertility.

Reduced soil fertility

Salinised (salt-affected) soils are typically found in marine-derived sediments, coastal locations and hot arid areas where capillary action brings salts to the upper part of the soil. Soil salinity has been a major problem in Australia following the removal of vegetation in dry land farming. Atmospheric deposition of heavy metals and persistent organic pollutants can make soils less suitable for sustaining the original land cover and land use.

Desertification (enlargement of deserts – see below) can be associated with this degradation. Climate change will probably intensify the problem, as it is likely to affect hydrology and hence land use.

Land degradation

Land degradation (Figure 5.8) is a decline in land quality and its productivity.

Overgrazing and agricultural mismanagement affect more than 12 million km^2 worldwide. 20% of the world's pastures and rangelands have been damaged and the situation is most severe in Africa and Asia. Huge areas of forest are cleared for logging, fuelwood, farming or other human uses.

Desertification

Desertification (Figure 5.9) can be natural, but increasingly it is the result of human activities:

- It occurs in rich countries such as the USA, Australia and Spain, as well as poor countries such as Burkina Faso, Mali and Ethiopia.
- Changes in agriculture, such as the concentration of livestock herds near boreholes, are partly responsible for desertification.
- It has major social, economic and environmental impacts.
- Social impacts can include increased hunger, reduced performance at school and increased illness.
- Economic impacts include reduced crop yields, falling incomes and reduced ability to work.
- Environmental impacts include reduced soil cover, decreased soil organic content, reductions in soil moisture availability and increased potential for soil erosion.

Expert tip

Make sure you study the textural groups. A 'clay' soil might only have 40% clay, and a mixed soil (loam) might only have as little as 10% clay. Some of the textural classes (groups) contain a wide variation of percentages.

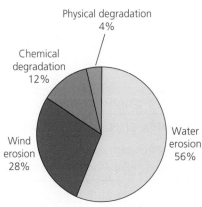

Figure 5.7 Types of soil degradation

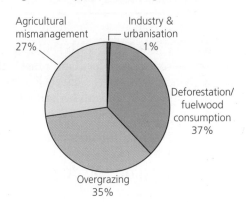

Figure 5.8 Types of land degradation

Keyword definition

Desertification – The spread of desert-like conditions into previously green areas, causing a long-term decline in biological productivity.

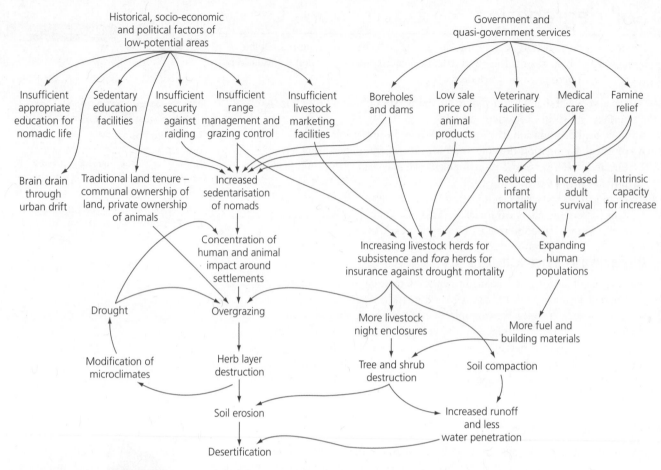

Figure 5.9 Some causes of desertification

■ **QUICK CHECK QUESTIONS**

14 What proportion of soil degradation is caused by water and wind erosion?

15 Suggest *two* economic and *two* social impacts of desertification.

Soil conservation measures

Revised ☐

Soil conservation measures include:

- abatement techniques such as afforestation
- soil conditioners (for example, use of lime and organic materials)
- wind reduction techniques (wind breaks, shelter belts, strip cultivation)
- cultivation techniques (terracing, contour ploughing)
- efforts to stop ploughing of marginal lands.

Many of these are used in combination.

■ Managing soil degradation

- Abatement strategies for combating accelerated soil erosion, such as afforestation, are lacking in many areas.
- To reduce the risk of soil erosion, farmers are encouraged towards more extensive management practices such as organic farming, afforestation, pasture extension and benign crop production.
- Methods to reduce or prevent erosion can be mechanical, including physical barriers such as embankments and wind breaks, or they can focus on vegetation cover and soil husbandry.
- More vegetation cover increases infiltration and reduces overland flow.

◼ Mechanical methods

- ◾ Mechanical methods include bunding, terracing, contour ploughing and shelterbelts such as trees or hedgerows.
- ◾ The key is to prevent or slow the movement of rainwater downslope.
- ◾ Contour ploughing forms at right angles to the slope to prevent or slow the downward accretion of soil and water.
- ◾ On steep slopes and in areas with heavy rainfall, such as the monsoon in Southeast Asia, contour ploughing is insufficient and terracing is undertaken. The slope is broken up into a series of flat steps, with bunds (raised levées) at the edge.
- ◾ The use of terracing allows areas to be cultivated that would not otherwise be suitable.
- ◾ In areas where wind erosion is a problem shelterbelts of trees or hedgerows are used.
- ◾ The trees act as a barrier to the wind and disturb its flow. Wind speeds are lowered, which reduces its ability to disturb the topsoil and erode particles.

◼ Cropping techniques

Preventing erosion by different cropping techniques focuses largely on:

- ◾ maintaining crop cover for as long as possible
- ◾ keeping in place the stubble and root structure of the crop after harvesting
- ◾ planting a grass crop – grass roots bind the soil, minimising the action of the wind and rain on a bare soil surface. Increased organic content allows the soil to hold more water, thus preventing aerial erosion and stabilising the soil structure.

◼ Salt-affected soils

There are three main approaches in the management of salt-affected soils:

- ◾ flushing the soil and leaching the salt away
- ◾ application of chemicals, for example gypsum (calcium sulfate) to replace the sodium ions on the clay and colloids with calcium ions
- ◾ a reduction in evaporation losses to reduce the upward movement of water in the soil.

Equally specialist methods are needed to decontaminate land made toxic by chemical degradation (see also eutrophication on pages 100–103).

◼ Soil management strategies in subsistence and commercial farming systems

Shifting agriculture in tropical rainforest is an example of subsistence farming. Soils in the tropical rainforest are very infertile. The high temperatures and high rainfall produce deeply weathered and leached soils that are lacking in nutrients.

The long growing season and intense competition among plants mean that most of the available nutrients are held by the biomass (trees). To increase the fertility of the soil, cultivators cut down ('slash') the vegetation and burn it, thereby releasing some of the nutrients into the soil. This increases the soil fertility in the short-term and allows agriculture to take place.

Over the following few years, these nutrients are washed away and the soil fertility drops. This forces the cultivators to abandon the plot they are farming and move to another plot – hence the term 'shifting cultivation'.

Some researchers believe that shifting cultivation may lead to long-term decline in soil fertility rather than being sustainable (Figure 5.10).

Common mistake

Some students make statements such as 'desertification occurs in LEDCs due to poor farming techniques'. This may be true but desertification also occurs in MEDCs, and may be the result of natural factors, such as long-term climate change.

Expert tip

There are many types and causes of soil degradation. Make sure that you can expand on the types of soil degradation (e.g. the different types of erosion or chemical degradation).

◼ QUICK CHECK QUESTIONS

16 Identify *four* types of soil conservation.

17 Briefly explain how afforestation acts as a measure of soil conservation.

18 State *two* soil conservation measures that could be taken to prevent desertification.

Expert tip

Attempts to achieve soil conservation can be affected by population growth and the need to produce more food, climate change and poverty-reduction schemes.

Common mistake

Many students seem to think that soil degradation only occurs in LEDCs. It occurs in MEDCs too. A classic example from the USA is the Dust Bowl of the 1930s.

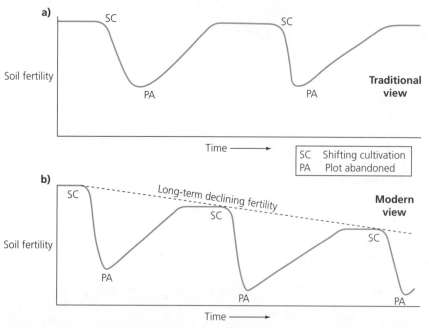

Figure 5.10 Effects of shifting cultivation on soil fertility

---

Common mistake

The tropical rainforest has been described as 'a desert covered by trees'. There is a paradox in that some of the world's most luxuriant vegetation is found on some of the world's least fertile soils. This is explained by the way in which most of the nutrients are held by the trees.

CASE STUDY

SOIL CONSERVATION ON THE GREAT PLAINS OF THE USA

The Great Plains of the USA experienced severe droughts in the 1930s, and the soils suffered severe wind erosion. Soil conservation techniques included:

- contour ploughing
- strip cultivation with an alternation of cultivated and fallow (crop-free) land
- temporary cover crops, such as a fast-growing millet
- shallow ploughing to eliminate weeds and conserve crop residues on the surface
- summer fallow to regenerate soil nitrogen as well as conserve moisture
- some areas being converted to permanent grazing
- use of a 'grass-break' to help stabilise soils by the accumulation of organic matter.

Since the end of the Second World War, the introduction of herbicides has made weed control possible.

Overall, these measures have lowered the risk of soil erosion and reduced nitrogen loss.

Expert tip

You should be able to:

- **explain** the relationship between soil ecosystem succession and soil fertility
- **discuss** the influences of human activities on soil fertility and soil erosion
- **evaluate** the soil management strategies of a given commercial farming system and of a given subsistence farming system.

■ QUICK CHECK QUESTIONS

19 In what ways is shifting cultivation a response to the soils of the tropical rainforest?

20 Identify *three* forms of soil conservation that were used on the Great Plains.

EXAM PRACTICE

3 Explain the importance of soil organisms in ecosystems. [5]

4 Outline how soil degradation can be caused by human activities. [5]

Topic 6 Atmospheric systems and societies

6.1 Introduction to the atmosphere

> **SIGNIFICANT IDEAS**
> - The atmosphere is a dynamic system that is essential to life on Earth.
> - The behaviour, structure and composition of the atmosphere influence variations in all ecosystems.

The atmosphere is a dynamic system

- The atmosphere is a dynamic system (with inputs, outputs, flows and storages) that has undergone changes throughout geological time.
- The atmosphere is a mixture of solids, liquids and gases that are held to the Earth by gravitational force.

■ Atmospheric composition

Up to a height of around 80 km the atmosphere is fairly similar, consisting of nitrogen (78%), oxygen (21%), argon (0.9%) and a variety of other trace gases such as carbon dioxide, helium and ozone. In addition it contains water vapour and solids (aerosols), such as dust, ash and soot.

There is no outer limit for the atmosphere, but most 'weather' occurs in the lowest 15 km – the troposphere.

■ Human impact on the atmosphere

Human activities impact atmospheric composition through altering inputs and outputs of the system. Changes in the concentrations of atmospheric gases – such as ozone, carbon dioxide and water vapour – have significant effects on ecosystems.

■ Atmospheric reactions

Most reactions connected to living systems occur in the inner layers of the atmosphere (Figure 6.1), which are the troposphere (around 0–15 km above sea level) and the stratosphere (around 10–50 km above sea level).

■ Clouds

Most clouds form in the troposphere and play an important role in the **albedo** effect of the planet. Albedo varies from 0.08 for moist, dark soils to about 0.9 for fresh snow. Clouds have an average albedo of about 0.5, so greatly restrict the amount of short-wave radiation reaching the ground below them. Low, thick clouds, such as stratus clouds (Figure 6.2), reflect more short-wave radiation than thin, wispy, cirrus clouds.

> **Keyword definition**
> **Albedo** – A measure of the reflecting power of a surface in relation to the amount of short-wave radiation received.

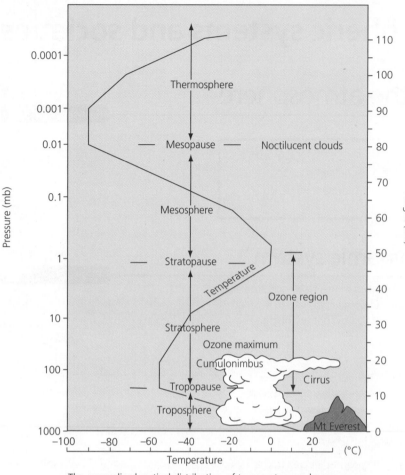

Figure 6.1 The structure of the atmosphere

The generalised vertical distribution of temperature and pressure up to about 110 km. Note particularly the tropopause and the zone of maximum ozone concentration with the warm layer above.

Figure 6.2 Stratus clouds

The greenhouse effect

Revised ▢

The greenhouse effect of the atmosphere (see page 135) is a natural and necessary phenomenon maintaining suitable temperatures for living systems.

- In the troposphere, temperatures fall with height.
- Certain gases are concentrated at certain heights.
- Ozone occurs in the stratosphere, mostly at around 25–30 km.

Expert tip

You should be able to:

- **discuss** the role of the albedo effect from clouds in regulating global average temperature
- **outline** the role of the greenhouse effect in regulating temperature on Earth.

Common mistake

The greenhouse effect is a natural process and is necessary for life on Earth. The accelerated or enhanced greenhouse effect refers to the changes in the greenhouse effect that are commonly referred to as global warming.

Expert tip

You are not required to memorise chemical equations for atmospheric gases.

■ QUICK CHECK QUESTIONS

1 Identify the two main gases in the troposphere.
2 In which part of the atmosphere does most weather occur?

6.2 Stratospheric ozone

> **SIGNIFICANT IDEAS**
> - Stratospheric ozone is a key component of the atmospheric system because it protects living systems from the negative effects of ultraviolet radiation from the Sun.
> - Human activities have disturbed the dynamic equilibrium of stratospheric ozone formation.
> - Pollution management strategies are being employed to conserve stratospheric ozone.

The role of ozone

- Certain gases are concentrated at certain heights.
- Ozone occurs in the stratosphere (Figure 6.1), mostly at around 25–30 km, and is important for the filtering of harmful ultraviolet radiation.

Ultraviolet radiation is absorbed during the formation and destruction of ozone from oxygen:

- Ozone absorbs ultraviolet radiation.
- It also absorbs some out-going long-wave radiation, so it is a greenhouse gas too.
- Ozone is created by oxygen rising up from the top of the troposphere and reacting with sunlight.
- Short-wave (ultraviolet) radiation breaks down oxygen molecules into two separate oxygen atoms.
- The oxygen atoms (O) combine with oxygen molecules (O_2) to form ozone (O_3).
- Stratospheric ozone is the high-level ozone that protects us from ultraviolet radiation.
- Tropospheric ozone or ground-level ozone is the 'bad' ozone that causes poor air quality and respiratory problems.
- Ozone can be destroyed naturally. In winter, clouds of ice particles form in the upper atmosphere. Chemical reactions take place on the ice involving chlorine. This can destroy ozone. By the summer, the ice clouds have disappeared and the destruction of ozone ceases.
- Human activities can also destroy ozone.

> **Expert tip**
>
> You are not required to memorise chemical equations.

> **Common mistake**
>
> There are two types of ozone – stratospheric and tropospheric. Stratospheric ozone is good – it protects against ultraviolet radiation – whereas tropospheric ozone is bad – it is a pollutant.

> **■ QUICK CHECK QUESTIONS**
>
> 3 In which part of the atmosphere is the ozone that protects against ultraviolet radiation found?
>
> 4 In which part of the atmosphere is the ozone that is damaging to human health found?

Ozone-depleting substances (ODSs)

The chemicals that cause stratospheric ozone depletion include halocarbons, such as chlorofluorocarbons (CFCs), hydrochlorofluorocarbons (HCFCs), methyl bromide, bromine and halons. These are found in refrigerators, air conditioners, aerosols, foamed plastics, pesticides, fire extinguishers and solvents.

Halogenated organic gases are very stable under normal conditions but can liberate halogen atoms when exposed to ultraviolet radiation in the stratosphere. These atoms react with monatomic oxygen and slow the rate of ozone reformation.

Pollutants enhance the destruction of ozone, thereby disturbing the equilibrium of the ozone production system. These pollutants cause 'holes' in the ozone layer. This in turn lets through ultraviolet radiation, which can be very damaging.

There is a clear seasonal pattern to the concentration of ozone. Each spring time there is a marked reduction in the amount of ozone over Antarctica. As summer develops, the ozone layer recovers. This is because in winter air over Antarctica becomes cut off from the rest of the atmosphere.

> **Keyword definition**
>
> **Halogenated organic gases** – Usually known as halocarbons, these were first identified as depleting the ozone layer in the stratosphere. They are now known to be potent greenhouse gases. The best-known are the chlorofluorocarbons (CFCs).

The intense cold allows the formation of clouds of ice particles, upon which chemical reactions involving chlorine can take place. In spring the chlorine atoms destroy ozone but by summer the ice clouds have disappeared and there is less destruction of ozone.

The effects of ultraviolet radiation

- Increased ultraviolet radiation is damaging to ecosystems that contribute significantly to global biodiversity, by damaging plant tissues, and causing the death of primary producers.
- It damages marine phytoplankton, which are some of the major primary producers of the biosphere.
- It is damaging to human populations around the world.
- It contributes to increased health costs.
- There are social and economic impacts of skin cancer and eye damage (cataracts).
- It can cause genetic mutations in DNA.

Ultraviolet radiation and aquatic food webs

The effects of ultraviolet radiation cause damage to photosynthetic organisms – especially phytoplankton – and their consumers, such as zooplankton. They cause reduced rates of photosynthesis.

Pollution management strategies

Methods of reducing the manufacture and release of ozone-depleting substances (ODSs) include the following:

- Fridges with ODS refrigeration can be replaced with 'green freeze' technology that uses propane and/or butane.
- Pump-action sprays can be used instead of aerosols.
- Alternatives to aerosols can be used – for example, soap rather than shaving foam.
- Organic methods of pest control can be used instead of methyl bromide.
- Old CFC coolants in fridges and air conditioning units can be recycled.

To protect against excessive ultraviolet radiation people can wear sunglasses, use sun block (sun cream), wear T-shirts and stay inside during the hottest part of the day.

> ### ■ QUICK CHECK QUESTIONS
> 5 State what is meant by the term *ozone hole*.
> 6 Identify *two* effects of increased ultraviolet radiation.

The role of UNEP

UNEP has a key role in providing information, and creating and evaluating international agreements for the protection of stratospheric ozone. In 1987 UNEP brought together 24 countries to sign the initial Montreal Protocol on Substances that deplete the Ozone Layer. Now 197 countries have signed the protocol. Production of ODSs fell from 1.8 million tonnes in 1987 to 45 000 tonnes in 2010. UNEP hopes to end production of HCFCs by 2040.

Illegal market for ODSs

An illegal market for ODSs persists and requires consistent monitoring.

The magnitude of the illegal trade in ODSs is between 7000 and 14 000 tonnes of CFCs annually. India and the Republic of Korea account for approximately 70% of the total global production of CFCs. Countries in the region with a high consumption of CFCs include the Philippines, Pakistan, Malaysia, India and China.

Reasons for the illegal trade in ODSs are many:

- ODS substitutes are often costlier than CFCs.
- Updating equipment to enable use of alternative chemicals is generally expensive.
- Penalties in many countries for ODS smuggling are small.

■ The Montreal Protocol

The 1987 Montreal Protocol on Substances that Deplete the Ozone Layer is the most significant and successful international agreement relating to an environmental issue, with many governments signing up and implementing the agreed changes. Subsequent revisions have reduced the phasing-out timescale because of the Protocol's success – phase out in Europe was achieved by 2000, and global phase out is expected by 2030.

The Protocol raised public awareness of the use of CFCs and provided an incentive for countries to find alternatives. Technology has been transferred to LEDCs to allow them to replace ozone-depleting substances, but some of the substances used are still ozone depleting, for example HCFCs. Some HCFCs have been replaced by HFCs (hydrofluorocarbons), but these are powerful greenhouse gases.

In spite of the success, there are issues that are hard to overcome:

■ It is harder for LEDCs to implement changes.
■ The second-hand appliance market means that old fridges are still in circulation.
■ It is a protocol that depends on national governments being willing to comply.
■ The long life of the chemicals in the atmosphere means that damage will continue for some time – until 2100.

■ QUICK CHECK QUESTIONS

7   Identify ways in which it is possible to reduce the impact of ozone depletion on human health.
8   Identify the ODS management strategies that attempt to deal with the causes of the pollution.

EXAM PRACTICE

1   Compare technocentric and ecocentric approaches to the management of ozone depletion.   [6]

Expert tip

Although the ozone 'holes' are located over Antarctica and the Arctic, the impacts of ozone depletion are worldwide.

Common mistake

The ozone 'hole' is not actually a hole but is a thinning of the concentration of ozone in the stratosphere.

Common mistake

Some students confuse global warming and ozone depletion by describing the Kyoto Protocol as relating to ozone depletion – this is incorrect. The Montreal Protocol relates to ODSs.

Expert tip

Remember to relate the potential solutions to ODSs to the causes, release of pollutants and impacts, as shown in Figure 4.9 on page 102.

Expert tip

You should be able to **evaluate** the role of national and international organisations in reducing emissions of ozone-depleting substances.

6.3 Photochemical smog

Revised ☐

SIGNIFICANT IDEAS

• The combustion of fossil fuels produces primary pollutants that may generate secondary pollutants and lead to photochemical smog, the levels of which can vary by topography, population density and climate.
• Photochemical smog has significant impacts on societies and living systems.
• Photochemical smog can be reduced by decreasing human reliance on fossil fuels.

Primary and secondary pollutants

Revised ☐

■ Primary pollutants from the combustion of fossil fuels include carbon monoxide, carbon dioxide, soot (black carbon), oxides of nitrogen, oxides of sulfur and unburned hydrocarbons.
■ Photochemical **smog** is a mixture of about 100 primary and secondary pollutants formed under the influence of sunlight. Ozone is the main pollutant.
■ Fossil fuels are burned and nitrogen oxides are released in vehicle emissions. In the presence of sunlight, these pollutants interact with others (e.g. volatile organic compounds) to produce secondary pollutants such as (tropospheric) ozone.
■ Tropospheric or ground-level ozone is a secondary pollutant because it is formed by reactions involving oxides of nitrogen (NOx).

Keyword definition

Smog – The term now used for any haziness in the atmosphere caused by air pollutants. Photochemical smog is produced through the effect of ultraviolet light on the products of internal combustion engines. It may contain ozone and is damaging to the human respiratory system and eyes.

- Ozone formation can take a number of hours, hence the polluted air may have drifted into suburban and surrounding areas.
- Smog is more likely under high-pressure (calm) conditions. Rain cleans the air and winds disperse the smog – these effects are associated with low-pressure conditions.
- The frequency and severity of photochemical smog in an area depend on local topography, climate, population density and fossil fuel use.

Tropospheric ozone

- When fossil fuels are burned, two of the pollutants emitted are hydrocarbons (from unburned fuel) and nitrogen monoxide (NO).
- Nitrogen monoxide reacts with oxygen to form nitrogen dioxide (NO_2), a brown gas that contributes to urban haze.
- Nitrogen dioxide can also absorb sunlight and break up to release oxygen atoms that combine with oxygen in the air to form ozone.
- The main source of NOx is road transport.

■ The effects of tropospheric ozone

- Ozone is a toxic gas and an oxidising agent.
- It damages crops and forests, irritates eyes, can cause breathing difficulties in humans and may increase susceptibility to infection.
- Ground-level ozone reduces plant photosynthesis and can reduce crop yields significantly.
- Ozone pollution has been suggested as a possible cause of the dieback of German forests (previously it was believed these had died as a result of acidification).
- It is highly reactive and can attack fabrics and rubber materials.

■ The effects of topography and temperature inversions

Urban microclimates also affect the production of ground level ozone. Urban areas generally have less vegetation than surrounding rural areas, and the concentration of buildings, industries and offices generates much heat.

Thermal (temperature) inversions trap the smog in valleys and basins – as in the cases of Los Angeles, Santiago, Mexico City, Rio de Janeiro, São Paulo and Beijing. The air is unable to disperse because cold air from the surrounding mountains and hills prevents the warm air from rising (Figure 6.3). Cold air is denser than warm air and so traps the warm air below.

Concentrations of air pollutants can build to harmful and even lethal levels, producing toxic and carcinogenic chemicals.

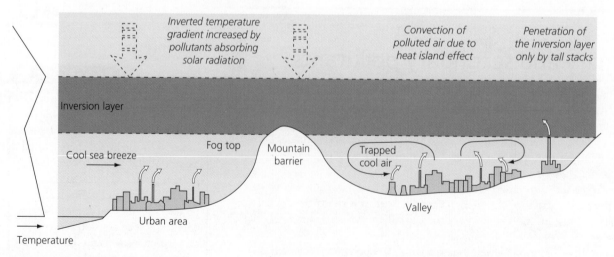

Figure 6.3 Temperature inversion

Deforestation and burning can also contribute to smog. The forest fires of Southeast Asia in the late 1990s and in 2012–2014 produced the Asian 'brown haze', throughout much of Malaysia, Indonesia and Singapore.

■ QUICK CHECK QUESTIONS

9 Explain what is meant by the term *temperature inversion*.

10 Why is ozone a problem in areas experiencing atmospheric high pressure?

■ Economic losses

Economic losses include the cost of clean-up strategies, loss of tourism, decreased worker productivity, increased cost of health care, decreased crop productivity and the cost of replacing materials. The World Bank indicates that the cost of air pollution in China is worth approximately 4% of its GDP annually.

Pollution management strategies

Revised ☐

There are many management strategies for tackling urban air pollution:

- Reducing fossil fuel combustion is an effective way of limiting the release of pollutants.
- Increased use of public transport can reduce total emissions of fossil fuels.
- Promotion of clean technology/hybrid cars.
- Provision of park-and-ride schemes to limit the number of cars entering urban areas.
- Preventing cars from parts of the city, i.e. pedestrianising part of the centre (e.g. Curitiba, Brazil).
- Only allowing certain cars into a city – for example those with odd-numbered registration plates on certain days, even numbered ones on other days (e.g. Mexico City).
- Fitting cars with catalytic convertors to reduce emissions of NOx.
- Reducing fossil fuel combustion by switching to renewable energy methods.
- Reducing fossil fuel combustion through urban design (e.g. south-facing windows, triple-glazed windows, cavity and loft insulation).
- Relocating industries and power stations away from centres of population.
- Ensuring industries and power stations have tall chimneys to help disperse pollutants.
- Filtering and catching pollutants at the point of emission.
- Designing cities so that there are more open spaces and water courses to help reduce the temperature and allow evaporative cooling (e.g. the restoration of the Cheong-Gye-Cheon River in Seoul, South Korea).
- Wearing masks to reduce inhalation of pollutants.

However:

- Most urban pollution comes from cars, especially old cars.
- Vehicles using diesel produce emissions of particulate matter.
- Catalytic convertors reduce fuel efficiency and increase CO_2 emissions.
- Public transport can be expensive and may be inconvenient.
- Sustainable urban design is expensive.

> **Expert tip**
>
> Larger cities with more vehicles will produce more tropospheric ozone than smaller cities.

> **Common mistake**
>
> Areas suffering the worst effects of ozone are not necessarily the areas that produced the pollution – ozone-polluted air can drift into suburban areas as it takes a number of hours to form in sunlight.

> **Expert tip**
>
> You should be able to **evaluate** pollution management strategies for reducing photochemical smog.

■ QUICK CHECK QUESTIONS

11 Suggest how cars can be prevented from entering a city centre.

12 Suggest how urban design can help reduce the incidence of petrochemical smogs.

> **EXAM PRACTICE**
>
> **2** Discuss, with reference to examples, the human factors that affect the successful implementation of pollution management strategies. [6]

6.4 Acid deposition

SIGNIFICANT IDEAS

- Acid deposition can impact living systems and the built environment.
- The pollution management of acid deposition often involves cross-border issues.

The causes of acid deposition

Rain is naturally acidic – because of carbon dioxide in the atmosphere – with a pH of about 5.6. Acid rain – or, more precisely, acid deposition – is the increased acidity of rainfall and dry deposition, as a result of human activity. The major causes of acid rain are the sulfur dioxide and nitrogen oxides produced when fossil fuels such as coal, oil and gas are burned. Sulfur dioxide and nitrogen oxides are released into the atmosphere where they can be absorbed by the moisture and become weak sulfuric and nitric acids, sometimes with a pH of around 3.

Dry deposition typically occurs close to the source of emission and causes damage to buildings and structures (Figure 6.4). **Wet deposition**, by contrast, occurs when the acids are dissolved in precipitation, and can fall at great distances from the sources. Wet deposition is an example of 'trans-frontier' pollution, as it crosses international boundaries.

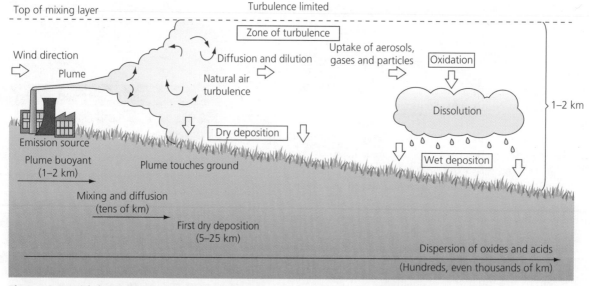

Figure 6.4 Acid deposition

Common mistake

Many students suggest that catalytic convertors are an example of how to tackle emissions of NOx, which is correct, but they fail to mention that they increase CO_2 emissions and reduce fuel efficiency.

Expert tip

You should consider measures to reduce fossil fuel combustion – for example, reducing demand for electricity and private cars and switching to renewable energy. Refer also to clean-up measures such as catalytic converters.

Expert tip

Knowledge of chemical equations is not required.

The effects of acid deposition

Acidic rainfall has a major impact on coniferous trees (Figure 6.5). Coniferous trees are more at risk than deciduous trees because they do not shed their leaves at the end of the year. Trees can also take up toxic aluminium ions from the soil.

In soils, increasing acidity leads to falling numbers of fungi, bacteria and earthworms. Earthworms cannot tolerate soils with a pH of below 4.5.

With increased acidity, more calcium and magnesium can be leached from a soil, while other metals, especially iron and aluminium, are mobilised by acidic water and flushed into streams and lakes. Aluminium damages fish gills, causing mucus to build up, thereby making gas exchange difficult.

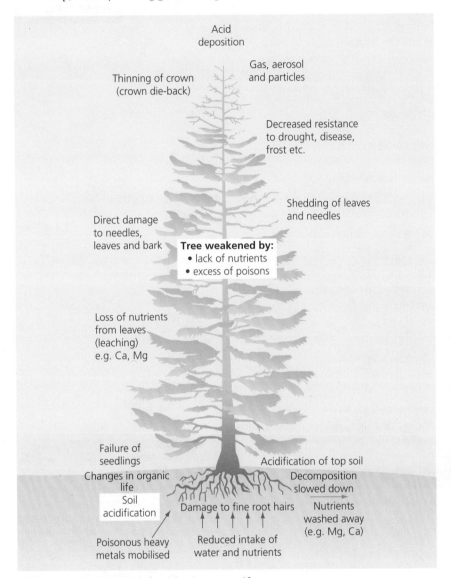

Figure 6.5 The impact of acid rain on coniferous trees

■ QUICK CHECK QUESTIONS

13 Identify the *two* chemicals associated with acid deposition.

14 Describe *one* impact of acidification on coniferous trees and *one* impact on soils, as shown in Figure 6.5.

Common mistake

Many students forget about dry deposition when referring to acid deposition. Both wet and dry deposition are important – dry deposition tends to be local, whereas wet deposition is more regional in scale.

■ Regional impacts of acidification

Figure 6.4 showed that dry deposition typically occurs close to the source of emission and causes damage to buildings and structures, whereas wet deposition occurs when the acids are dissolved in precipitation, and may fall at great distances from the sources.

The main areas experiencing acid rain are those areas downwind of major industrial regions such as Scandinavia (downwind from Western Europe), northeast USA and eastern Canada (downwind from the US industrial belt). There is less acidification in Scandinavia now compared with the 1980s, as there is less heavy industry in Western Europe. Areas that are increasingly causing acidification include China and India.

Areas experiencing acidification usually have a combination of being downwind from industrial belts and fossil fuel power stations, having high rainfall, containing lots of forests and lakes, and having thin soils.

Some environments are able to neutralise the effects of acid rain. This is referred to as their **buffering capacity**. For example, chalk and limestone areas are very alkaline and can neutralise acids very effectively. The underlying rocks over much of Scandinavia, Scotland and northern Canada are granite. These are naturally acid, and have a very low buffering capacity. It is in these areas that there is the worst damage from acid rain.

Pollution management strategies

Revised ☐

Various methods are used to try to reduce the damaging effects of acid deposition. One of these is to add powdered limestone to lakes to increase their pH values. However, the only really effective and practical long-term treatment is to curb the emissions of the offending gases. This can be achieved in a variety of ways:

- by reducing the demand for electricity
- by reducing the amount of fossil fuel combustion
- use of low-sulfur fossil fuels
- use of alternative energy sources that do not produce nitrate or sulfate gases (e.g. hydropower or nuclear power)
- by removing the pollutants before they reach the atmosphere
- use of limestone scrubbers in chimneys of power stations (to neutralise the acid)
- spraying powdered limestone onto acidified soils or waters.

There have been various international agreements concerning acidification:

- The 1979 Convention on Long-Range Transboundary Air Pollution was important for the clean-up of acid rain in Europe.
- The 1999 Gothenburg Protocol commits countries to reducing emissions of sulfur dioxide, oxides of nitrogen, VOCs and ammonia in an attempt to reduce acidification, eutrophication and ground-level ozone.
- The 1991 Air Quality Agreement between the USA and Canada focused on acid rain and, later, smog.

■ Clean-up and restoration

Although there are clean-up and restoration measures, such as the use of powdered limestone in acidified lakes and soils, and the recolonisation of degraded ecosystems, the scope of these measures is limited.

Expert tip

The case against SO_2 and NOx is not clear-cut. While victims and environmentalists stress the risks of acidification, industrialists stress the uncertainties. For example:

- Rainfall is naturally acidic and could cause some of the damage.
- No single industry/country is the sole emitter of SO_2/NOx – so it is impossible to apportion blame.
- Car owners with catalytic convertors reduce emissions of NOx.
- Different types of coal have variable sulfur content – some coal is 'cleaner' than others.

Common mistake

Some students think that acidification is no longer a problem as less SO_2 is being emitted. Although less coal is being burned in some countries, the increase in the number of cars more than compensates for this.

Expert tip

You should be able to evaluate pollution management strategies for acid deposition.

EXAM PRACTICE

3 Evaluate the role of reducing, reusing and recycling strategies in the management of atmospheric pollutants. [9]

■ QUICK CHECK QUESTIONS

15 Identify the rocks that offer the best buffering against acidification.

16 Explain how adding powdered lime can combat acidification.

7.1 Energy choices and security

> **SIGNIFICANT IDEAS**
> * There is a range of different energy sources available to societies that vary in their sustainability, availability, cost and socio-political implications.
> * The choice of energy sources is controversial and complex. Energy security is an important factor in making energy choices.

Energy sources

Fossil fuels, such as coal, oil and natural gas, make up the majority of our energy supply; their use is expected to increase to meet global energy demand. Figure 7.1 shows how energy resources have changed over much of the twentieth century and how they are predicted to change up until 2050.

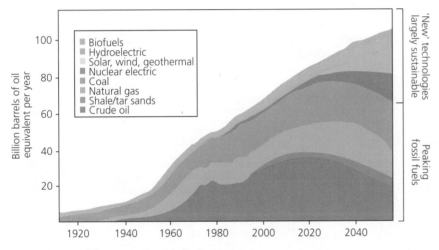

Legend:
- Biofuels
- Hydroelectric
- Solar, wind, geothermal
- Nuclear electric
- Coal
- Natural gas
- Shale/tar sands
- Crude oil

y-axis: Billion barrels of oil equivalent per year

x-axis: 1920 1940 1960 1980 2000 2020 2040

right labels: 'New' technologies largely sustainable / Peaking fossil fuels

Figure 7.1 World energy demand – long-term energy sources

Energy consumption has traditionally been much higher in rich countries (more economically developed countries) than poor countries (less economically developed countries). This is because rich countries have more energy sources and economies that are based on energy-intensive industries.

As poor countries industrialise (newly industrialising countries – NICs – such as China and India), they are using vastly more energy that they did a few decades ago.

> ■ **QUICK CHECK QUESTIONS**
> 1 What was/is the most common energy source:
>   a) in 1950
>   b) in 2000
>   c) projected for 2050?
> 2 Which energy sources are predicted to show the most growth by 2050 compared with 2000?
> 3 Which of these are **renewable** and which are **non-renewable**?

> **Expert tip**
>
> Figure 7.1 is an example of a compound line graph. This means that the value for any category is drawn above the category below it. For example, the amount of coal used in 2000 was about 20 billion barrels of oil equivalent, not 60 billion. Oil was approximately 30 billion barrels and natural gas 10 billion, and so coal is drawn between 40 and 60. The value of a compound graph like this is that it shows you total (consumption of energy) as well as the relative contribution, in this case, of each of the energy resources.

> **Keyword definitions**
>
> **Renewable** – Natural resources that have a sustainable yield or harvest equal to, or less than, their natural productivity – for example, timber.
>
> **Non-renewable** – Natural resources that cannot be replenished within a timescale of the same order as that at which they are taken from the environment and used – for example, fossil fuels.

The advantages and disadvantages of contrasting energy sources

> **Keyword definition**
>
> **Fossil fuels** – Non-renewable resources including oil, coal, natural gas and shale gas.

▓ Oil

Table 7.1 Advantages and disadvantages of oil

Advantages	Disadvantages
Oil has been a relatively cheap and efficient form of energy, and a versatile raw material	The importance of oil as the world's leading fuel has had many negative effects on the natural environment. For example:
It is relatively easy to transport by tanker or pipeline	• oil slicks from tankers (e.g. *Torrey Canyon* (1967), *Exxon Valdez* (1987) and *Braer* (1993)) and drilling rigs (e.g. Deepwater Horizon (2010))
Oil reserves are generally found in geological structures such as anticlines, fault traps and salt domes	• damage to coastlines, fish stocks and communities dependent upon the sea
At present rates of production and consumption reserves could last for another 40 years	• water pollution caused by tankers illegally washing/cleaning out tanks in the North Sea
Nearly two-thirds of the world's reserves are found in the Middle East	• Gulf War damage – storage of oil and oil wells can be targets for destruction, causing immeasurable environmental damage
The Arctic is believed to have about 35% of the world's oil reserves	It is a finite resource and will eventually run out – indeed we may already have experienced 'peak oil'
	The burning of oil contributes to global warming through the release of CO_2

▓ Hydroelectric power (HEP)

Table 7.2 Advantages and disadvantages of HEP

Advantages	Disadvantages
Hydroelectric power (HEP) is a renewable form of energy that harnesses fast-flowing water with a sufficient volume	HEP plants are very costly to build
It is considered to be a clean form of energy as it does not emit greenhouse gases (although many are released during the construction of the dam)	Only a small number of places have a sufficient head of water
HEP stations are often associated with aluminium smelters to use up excess energy	Markets are critical – the plant needs to run at full capacity to be economic
	Migratory fish and mammals may have their routes affected
	There may be increased evaporation behind the dam and the deposition of silt
	Diseases such as schistosomiasis can be spread by the stagnant water
	Fish yields downstream can be adversely affected by the trapping of sediments behind the dam

▓ Nuclear power

Table 7.3 Advantages and disadvantages of nuclear power

Advantages	Disadvantages
Nuclear power accounted for about 9% of energy production in 2005	Nuclear power stations are very expensive to build and have nearly always overrun on costs and time needed for completion
Nuclear energy has received support from the environmental movement in recent years; this has occurred for two reasons:	The decommissioning costs of obsolete nuclear power stations are enormous
• the prospect of global warming	There are serious risks related to radiation (e.g. Chernobyl, Fukushima); the 2011 Fukushima disaster in Japan led to Japan turning its back on nuclear power
• nuclear power does not emit greenhouse gases	

> **Expert tip**
>
> It is not just the non-renewable forms of energy that contribute to global warming. Renewable forms such as biomass (fuelwood and dung) release CO_2. Many other forms, including HEP, tidal and nuclear, contribute to global warming in the construction of, in this case, dams, barrages and power stations.

Coal

Table 7.4 Advantages and disadvantages of coal

Advantages	Disadvantages
Coal is mostly found in the mid-latitudes of the northern hemisphere and has been responsible for the development of western industrial growth	Coal is dirty, bulky, costly and difficult to transport
Thick, level, continuous seams are the most competitive and facilitate the use of machinery	Due to inefficiencies in early forms of transport and machinery, industries had to be located on the coalfields
The two principal users of coal as a fuel are production of electricity in thermal power stations and the smelting industry (e.g. iron and steel)	Coal has a negative impact upon the environment in a number of ways: • open-cast mining causes serious visual and noise pollution • burning coal contributes to acid rain and to global warming
Coal is also important in the chemicals industry and provides a range of products from aspirin to nylon	
Coal is sometimes divided into two main types, coking coal and steam coal: • coking coal is used in the manufacture of iron and steel • steam coal, however, is used to generate electricity; the demand for steam coal is rising rapidly	

Some of these disadvantages can be managed:

- The introduction of precipitators and filters in smokestacks retains dust, sulfur dioxide and nitrogen oxides (NOx) within the chimney.
- Sulfur emissions have declined by up to 95% and NOx emissions by 60%.

Solar power

Table 7.5 Advantages and disadvantages of solar power

Advantages	Disadvantages
No finite resources are involved	It is affected by cloud, seasons and daylength, so it cannot be guaranteed to work in all locations
Less environmental damage is caused	It is not always possible where demand exists
No atmospheric pollution is given off	
It is suitable for small-scale and large-scale production	The high costs of solar power make it difficult for the industry to achieve its full potential
Energy from the Sun is clean, renewable and so abundant that the amount of energy received by the Earth in 30 minutes is the equivalent to all the power used by humans in 1 year	At present it does not make a significant contribution to energy efficiency
	Each unit of electricity generated by solar energy costs between four and ten times as much as that derived from fossil fuels

> **Common mistake**
>
> Nuclear power is sometimes described as an 'alternative energy' along with renewable forms of energy such as HEP and wind. It is an alternative to fossil fuels as it does not release carbon dioxide but it is not a renewable form of energy – once uranium has been used it cannot be recycled or used again.

Wind power

Table 7.6 Advantages and disadvantages of wind power

Advantages	Disadvantages
Wind power is suitable for small-scale production	Visual impact – although some people like the appearance of wind turbines, many dislike them
The only requirement is an exposed site, such as a hillside, flat land or close to the coast, where winds are strong and reliable	They are noisy
There is no pollution of air, ground or water	They can injure migrating birds
No finite resources are involved	Winds may be unreliable
It reduces environmental damage elsewhere	Large-scale development is hampered by the high cost of development, the large number of turbines needed, and the high cost of new transmission grids
	Suitable locations for wind farms are normally quite distant from centres of demand

■ Tidal power

Table 7.7 Advantages and disadvantages of tidal power

Advantages	Disadvantages
Tidal power is a renewable, clean energy source	Large-scale production of tidal energy is limited for a number of reasons:
It requires a funnel-shaped estuary, free of other developments, with a large tidal range	• high cost of development
The River Rance in Brittany, France and the Bay of Fundy in Canada are good examples of where tidal power has been developed	• limited number of suitable sites
	• environmental damage to estuarine sites
	• long period of development
	• possible effects on ports and industries upstream

■ Fuelwood

Table 7.8 Advantages and disadvantages of fuelwood

Advantages	Disadvantages
Fuelwood refers to the use of trees and vegetation as an energy source	In many areas fuelwood availability is decreasing fast
Fuelwood is the most important source of energy in rural Africa, due to its availability – for example, in Tanzania it accounts for 90% of energy consumption, and in Zimbabwe 50%	As fuelwood becomes more scarce, women and children travel further to collect it and round trips of 10 km are not uncommon; this imposes an extra burden upon their health

> **Expert tip**
>
> Try to make sure that your answers are balanced. If you have to answer a question on the advantages and disadvantages of a form of energy, make sure that you cover both sides and try to give as many advantages as disadvantages. You can still form a conclusion and give your views, while also recognising that others may have different views.

> **Expert tip**
>
> In the exams, you only have to consider one non-renewable resource (fossil fuels or nuclear) and one renewable energy source. Revise the ones you have studied in class.

A decline in fuelwood does not just mean pressure on time and labour. Trees are a multipurpose resource. They:

- are used to build houses
- provide fencing
- contribute to food supply
- supply drugs and medicines
- prevent wind and water erosion
- act as a habitat for wild animals.

> ## ■ QUICK CHECK QUESTIONS
>
> 4 What are the advantages of fossil fuels?
>
> 5 What are the disadvantages of renewable forms of energy, such as HEP and wind?

Energy security

Revised

Energy security is uninterrupted availability of energy sources at an affordable price that provides countries with a degree of independence (Figure 7.2). However, there are wide variations in access to energy at a national level down to a household level (e.g. through poverty). A lack of energy can have a negative impact on social and economic development, and in extreme cases may lead to conflict.

> **Keyword definition**
>
> **Energy security** – Having an adequate, reliable and affordable supply of energy.

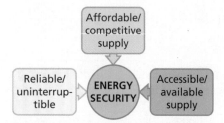

Figure 7.2 Defining energy security

Energy choices

Revised

There are many important factors to consider in the use of energy, relating to availability as well as economic, cultural, environmental and technological issues:

- Availability and reliability of supply – the UK used to have coal, then it had oil, but it has limited potential for solar or geothermal energy.
- Suitability and efficiency of supply – many poor countries are limited in their choice of energy source.
- Costs of production, distribution and use – is nuclear power or tidal energy too expensive?
- Type of market – industrial, agricultural or residential – energy demand soars as a country industrialises.
- Political factors – in 1973 OPEC (Organization of the Petroleum Exporting Countries) raised the price of oil, causing other countries to develop their own, cheaper resources.
- Demand for energy – this depends on a country's population size, wealth and level of industrialisation.
- Population growth – rapid population growth leads to more energy being used.
- Economic growth – rapid economic growth leads to more energy being used.
- Stage of development – less developed countries use a smaller amount of energy per head and more basic sources such as fuelwood, whereas developed countries use more energy and more expensive forms such as nuclear and oil.
- Climate factors – certain climates allow certain types of energy such as solar or wind power; colder climates require more heating.

Low levels of energy consumption per head in poor countries are explained by:

- lack of suitable resources
- lack of economic development to finance the rapid development of energy resources or imports of energy
- rapid growth of population (demand exceeds supply)
- lack of capital to develop alternative forms of energy
- debt
- lack of technological resources
- lack of trust, especially with regard to nuclear power
- lack of fuelwood.

Common mistake

Not all rich countries are the same – nor are all poor countries. It is very easy to over-generalise in an answer. Try to make sure that you refer to specific rich and poor countries and give some details about their energy consumption.

Expert tip

Make sure that you have two contrasting countries – a rich one and a poor one.

■ QUICK CHECK QUESTIONS

6 The diagram shows energy consumption in China by source. Describe the changes in China's energy sources, 1980–2005.

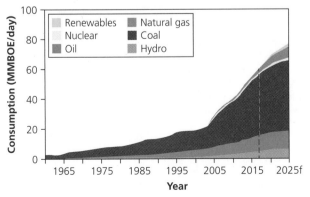

China's energy consumption from 1965 to 2025 (forecast)

7 The diagram shows China's energy consumption by economic sector.

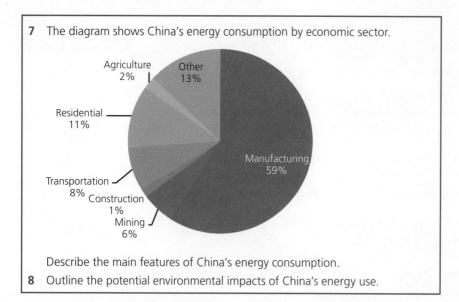

Describe the main features of China's energy consumption.

8 Outline the potential environmental impacts of China's energy use.

Energy efficiency and energy conservation

Energy efficiency refers to the attempts to improve products and services so that less energy is required for them to function. For example, fluorescent lights and LED lights reduce the amount of energy needed to provide light for an area. Many modern appliances, such as fridges, freezers and washing machines, require much less energy to function compared with older models.

Energy conservation refers to efforts to reduce energy consumption. For example, someone who wears extra clothing – or sits under a blanket – rather than turning the heating on is conserving energy. Having homes with south-facing windows (in the northern hemisphere) allows greater solar heating of rooms. Other methods of conservation include double- or triple-glazing of windows, cavity wall insulation, loft insulation, only filling a kettle with as much water as is needed rather than completely filling it and switching off lights when not in a room.

Expert tip

You should be able to:

• **evaluate** the advantages and disadvantages of different energy sources
• **discuss** the factors that affect the choice of energy sources adopted by different societies
• **discuss** the factors that affect energy security
• **evaluate** the energy strategy of a given society.

EXAM PRACTICE

1 Outline the advantages of *two* energy sources, and suggest why one plays a greater role in meeting energy needs than the other in a named country. [5]

2 Evaluate the advantages and disadvantages of *two* contrasting energy sources and discuss the economic factors that affect the choice of these energy sources by different named societies. [9]

7.2 Climate change – causes and impacts

SIGNIFICANT IDEAS

• Climate change has been a normal feature of the Earth's history, but human activity has contributed to recent changes.
• There has been significant debate about the causes of climate change.
• Climate change causes widespread and significant impacts on a global scale.

Climate and weather

Climate is the average and extreme of weather conditions over a period of not less than 30 years. It includes temperature, rainfall (all forms of precipitation), humidity, cloud cover, wind speed and direction, and air pressure. In contrast, the term weather refers to the state of the atmosphere at any given instant. Normally, we refer to the weather as being short-term, i.e. over a few days. Like climate, it includes temperature, rainfall (all forms of precipitation), humidity, cloud cover, wind speed and direction, and air pressure.

Atmospheric and oceanic circulatory systems

Weather and climate are affected by oceanic and atmospheric circulatory systems. On a global scale, the oceanic conveyor belt (Figure 7.3) transfers energy around, and links, the world's great oceans. Waters that flow from the Equator are generally warm, whereas those that flow from the high latitudes are cold.

> **Keyword definitions**
>
> **Maritime –** Relating to areas that are close to the coast.
>
> **Continental –** Relating to areas that are distant from the coast.

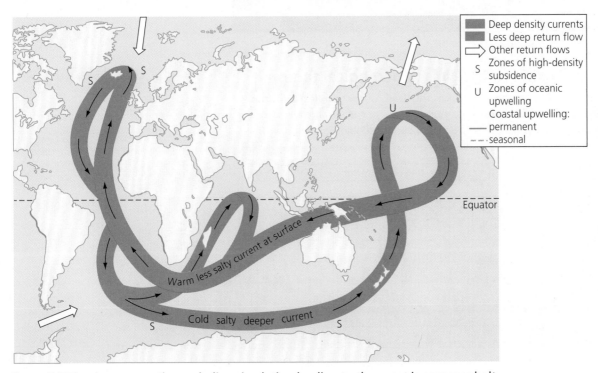

Figure 7.3 The deep ocean thermohaline circulation leading to the oceanic conveyor belt

Winds have a similar impact. The prevailing wind refers to the most frequent wind that an area receives. Winds from low latitudes tend to be warm and winds from high latitudes are cold. Winds that blow over oceans are moist whereas those that blow over land are dry. Some winds have a seasonal influence, such as the monsoon winds of Asia.

Greenhouse gases and the greenhouse effect

The greenhouse effect is a normal and necessary condition for life on Earth. **Greenhouse gases** allow incoming short-wave radiation to pass through the atmosphere and heat up the Earth's surface. They trap a proportion of the out-going long-wave radiation from the Earth – hence the atmosphere is heated from below rather than from above (Figure 7.4). In this way greenhouse gases raise the Earth's temperature by about 33°C and make life on Earth possible.

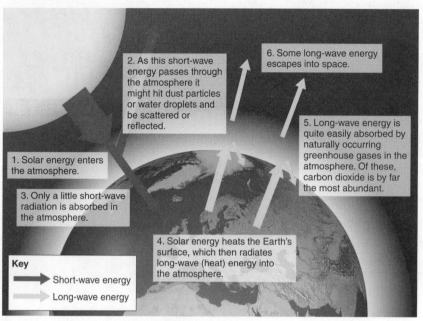

Figure 7.4 The greenhouse effect

■ Carbon dioxide (CO_2) levels in geological times

There have been considerable changes in the levels of carbon dioxide in the geological past (Figure 7.5). Generally, higher levels of carbon dioxide correlate with higher temperatures, and lower levels of carbon dioxide correlate with lower temperatures.

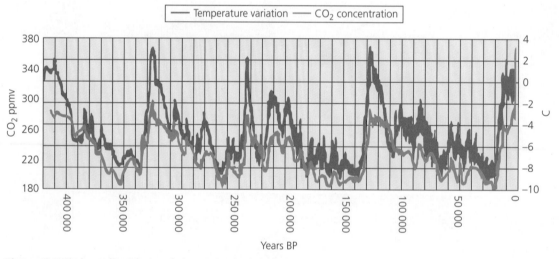

Figure 7.5 Carbon dioxide levels in geological time

Keyword definition
Greenhouse gases – Atmospheric gases that absorb infrared radiation, causing world temperatures to be warmer than they would otherwise be. This process is sometimes known as 'radiation trapping'. The natural greenhouse effect is caused mainly by water and carbon dioxide.

Human activities and greenhouse gases

The main greenhouse gases include water vapour, carbon dioxide, methane, CFCs, ozone and oxides of nitrogen. Human activities are increasing levels of carbon dioxide, methane and CFCs in the atmosphere, which may lead to **global warming**.

Common mistake

The greenhouse effect and global warming are often considered to be the same. They are not but they are related. The greenhouse effect is a natural process that traps some outgoing longwave radiation and enables life on Earth. Global warming – or the enhanced greenhouse effect – is the process in which human activities have led to an increase in the amount of greenhouse gases in the atmosphere, and an increased trapping of greenhouse gases, leading to an increase in global temperatures, i.e. global warming.

■ QUICK CHECK QUESTIONS

9 Briefly describe the greenhouse effect.

10 Describe the relationship between carbon dioxide and temperature as shown in Figure 7.5.

Revised ☐

Figure 7.6 The main greenhouse gas emissions related to human activity

Legend:
■ Carbon dioxide
■ CFCs, HFCs
■ Methane
■ Oxides of nitrogen

■ **QUICK CHECK QUESTIONS**

11 Identify the main source of human-related (anthropogenic) emissions of greenhouse gases.

12 List the anthropogenic greenhouse gases in descending order of emissions.

- Sources of carbon dioxide include respiration by living organisms, breakdown of organic material, volcanic vents, burning of fossil fuels and organic materials, and forest fires.
- There are many different sources of methane, including wetlands, bogs, stagnant water bodies, rice paddies, tundra soils, the breakdown of organic material, volcanic vents, livestock, landfill sites, melting of permafrost and manure or sewage.
- Sources of CFCs include refrigeration and air conditioning systems, plastic foams, aerosol cans and solvents used in the electronics industry.

Keyword definition

Global warming – An increase in average temperature of the Earth's atmosphere.

Common mistake

Many people forget that water vapour is a greenhouse gas, although it is not considered to be a human-related greenhouse gas. Its exact role is unclear.

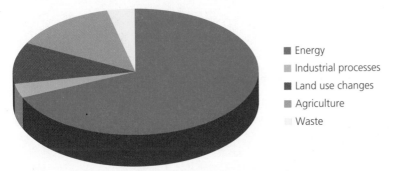

Figure 7.7 Sources of human-related greenhouse gases

Legend:
■ Energy
■ Industrial processes
■ Land use changes
■ Agriculture
■ Waste

Expert tip

Although there are some sources of carbon dioxide and methane that are the same, such as volcanic vents and the breakdown of organic matter, you need to give different sources rather than the same one for carbon dioxide and methane.

The impacts of climate change

Revised ☐

The potential effects on the distribution of biomes, global agriculture and human societies might be adverse or beneficial, for example:

- biomes shifting
- change in location of crop-growing areas
- changed weather patterns
- coastal inundation (due to thermal expansion of the oceans and melting of the polar ice caps)
- human health (spread of tropical diseases).

■ Changes in biotic components of ecosystems

Table 7.9 Changes in biotic components of ecosystems and their impact

Changes	Impacts/significance for humans
• Biomes shift to higher latitudes/altitudes as climate changes	• Shifting biomes mean crop-growing areas will shift
• Expansion of areas inhabited by tropical disease vectors such as mosquitoes	• Some areas will change in terms of productivity
• Loss of species diversity as species are unable to adapt or have limited scope for shifting, and become extinct	• A shortage of resources could lead to increased conflict, for example over water or food
• Animals can migrate but plants shift their range more slowly	• Impact on human health as more areas become affected by tropical diseases
• Increased rates of primary productivity	• Loss of economic, aesthetic and/or medical benefits of species diversity

Coastal inundation

Coastal inundation (flooding) will occur as global warming will lead to thermal expansion of water and melting of (land-based) glaciers and ice caps, leading to a sea-level rise. This could have many impacts:

- increased coastal erosion
- intrusion of salty water
- reduction of mangrove forests
- coral reefs unable to obtain sufficient light
- wading birds struggling to obtain food
- contamination of soils and a decline in agricultural production.

Impact on human health

Global warming could have a varied impact on human health:

- an increase in stagnant water, meaning more mosquitoes that carry diseases
- people forced to leave their homes and becoming more susceptible to diseases, especially children and the elderly
- changes in distributions of organisms, bringing new diseases to areas
- saltwater intrusion onto coastal agricultural land, meaning a reduction in food production, and leading to more widespread hunger and malnutrition, increasing the impact of diseases.

Changes in weather patterns

Researchers suggest that a doubling of CO_2 from the base line value of 270 ppm (parts per million) to 540 ppm would lead to:

- an increase in temperatures of around 2°C, although increased warming is likely to be greater at the poles rather than at the Equator
- changes in prevailing winds
- changes in precipitation
- continental areas becoming drier.

CASE STUDY

THE POTENTIAL IMPACTS OF CLIMATE CHANGE ON THE UK

Positive impacts might include:

- an increase in timber yields (up to 25% by the 2050s), especially in the north of the UK (with perhaps some decrease in the south)

- a northward shift of farming zones by about 200–300 km per degree centigrade of warming, or 50–80 km per decade, which would improve some forms of agriculture, especially pastoral farming in the northwest

- enhanced potential for tourism and recreation as a result of increased temperatures and reduced precipitation in the summer, especially in the south.

Negative impacts might include:

- an increase in droughts, soil erosion and the shrinkage of clay soils (Figure 7.8)

- an increase in invasive animal species – especially insects – as a result of northward migration from the continent, and a small decrease in the number of plant species due to the loss of northern and montane (mountain) species

- a decrease in crop yields in the southeast

- an increase in river flow in the winter and a decrease in the summer, especially in the south

- an increase in public and agricultural demand for water

Figure 7.8 One potential impact of global warming on the UK – drier soils

■ damage from increased storminess, flooding and erosion to natural and human resources and other assets in coastal areas

■ increased incidents of certain infectious diseases in humans and of the health effects of episodes of extreme temperature.

Table 7.10 What would be a significant climate change for the UK?

If there is an increase in temperature of 0.5°C	If there is an increase in temperature of 1°C	If there is an increase in temperature of 1.5°C
Summer and winter precipitation increases in the northwest by 2–3%	Summer and winter precipitation increases in the northwest by 4%; some precipitation decreases in the southeast by 5%	Summer and winter precipitation increases in the northwest by about 7%; summer precipitation decreases in the southeast by 7–8%
Annual runoff in the southern UK decreases by 5%	Annual runoff in the southern UK decreases by 10%	Annual runoff in the southern UK decreases by 15%
Frequency of 1995 type summer (drought) increases from 1:90 to 1:25	Frequency of 1995 type summer increases from 1:90 to 1:10	Frequency of 1995 type summer increases from 1:90 to 1:3
Disappearance from the British Isles of a few niche species, for example alpine wood fern and oak fern	Disappearance from the British Isles of certain species, for example ptarmigan and mountain hare	Further disappearance from the British Isles of several species
In-migration of some continental species and expansion of some native species, e.g. red admiral and painted lady butterflies, Dartford warbler	Expansion of range of most butterflies, moths and birds, such as goldeneye and redwing	In-migration of several species
Increase in overall UK timber productivity by 3%	Increase in overall UK timber productivity by 7%	Increase in overall UK timber productivity by 15%
Increase in demand for irrigation water by 21% over the increase without climate change, and in domestic demand by an additional 2%	Increase in demand for irrigation water by 42% over the increase without climate change, and in domestic demand by an additional 5%	Increase in demand for irrigation water by 63% over the increase without climate change, and in domestic demand by an additional 7%
Decrease in heating demand by 6%	Decrease in heating demand by 11%	Decrease in heating demand by 16%

■ QUICK CHECK QUESTIONS

13 State *one* advantage of rising temperatures for the UK.

14 Explain *two* reasons why global warming could lead to an increase in the number of cases of malaria.

Positive and negative feedbacks

Revised ▢

There are many examples of feedback in the process of global warming (Figures 7.9 and 7.10). Positive feedback mechanisms involve increasing temperatures, melting permafrost and the release of methane. As methane is a greenhouse gas, it has the potential to increase temperatures, thereby reinforcing the rise in temperature.

In contrast, higher temperatures in tropical areas may lead to increased precipitation in many parts of the world. Increased precipitation in polar areas leads to increased snow cover. The surface of snow and ice is very reflective, and so the albedo is increased. Increased reflectivity reduces the amount of solar radiation received and so lowers temperatures.

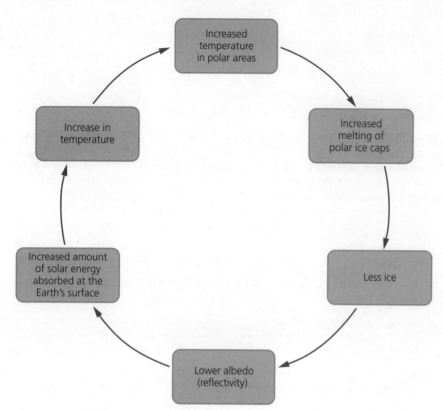

Figure 7.9 Positive feedback in global warming

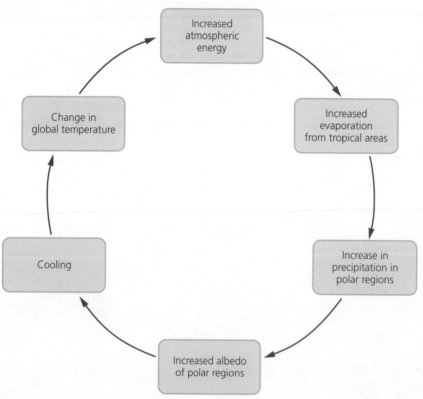

Figure 7.10 Negative feedback in global warming

▦ EVSs and climate change

There are many different viewpoints about global warming and these can change over time:

■ 350.org is an organisation that is building a global movement to solve the climate crisis.
■ Al Gore's book and film *An Inconvenient Truth* managed to get the message about global warming to many of the US public. However, some critics

argue that it was only the effects of Hurricane Katrina (and later Superstorm Sandy) that awoke many Americans to the idea that global warming was having an impact on the USA.

- In a similar way, some climate experts believe that the 2013 Australian heatwave will prove a turning point in how Australians respond to warnings about human-induced climate change. In a country that relies on fossil fuels for much of its wealth (coal is its second largest export and produces about 80% of its electricity), climate-change sceptics have often swayed political action.
- In *The Skeptical Environmentalist*, Bjorn Lomborg argues that there are too many uncertainties regarding global warming, although he accepts that human activity is adding to it.
- In *The Great Global Warming Swindle*, Martin Durkin argues that global warming is more likely to be caused by natural variations such as sunspot activity.

> ### ■ QUICK CHECK QUESTIONS
> 15 Identify the type of feedback that is self-regulating.
> 16 Identify the type of feedback that leads to instability and change.

> **Common mistake**
>
> Some students assume that negative feedback has a negative impact on the environment – this is not so, as negative feedback can be self-regulating and reduce the amount of change in an environment.

> **Expert tip**
>
> Be prepared to draw a diagram of positive and/or negative feedback in relation to global warming.

Complexity of climate change

Revised ☐

Some critics state that even if humans are causing climate change, the Earth will correct itself – this relates in part to the **Gaia hypothesis**, i.e. that the Earth is a self-regulating entity.

■ Those who believe that human-induced global warming is real

Many claim that the scientific data prove that the climate is warming. Data from a variety of sources show that carbon dioxide levels and greenhouse gas levels are increasing, and in some cases that temperatures are rising.

They claim that human activities and/or fossil fuel combustion are known to increase carbon dioxide/greenhouse gas levels, and insist that carbon dioxide and other greenhouse gases are known to affect global temperatures. Therefore it is likely that human activities are resulting in global climate change. Moreover, the rapid rate of increase in carbon dioxide levels implies a human link.

Some claim that natural fluctuations occur, so changes in climate could still be a short-term trend. They argue that the only technologically verifiable data have been collected from a short period of time.

They also state that other aspects of climate change are not all fully understood and that climate has changed in the past. This is in part due to natural fluctuations such as Milankovitch cycles (variations in the Earth's orbit around the Sun, in the length of seasons and in the orientation of the poles towards or away from the Sun).

Moreover, current carbon dioxide levels and global temperature fluctuations are moderate compared with geologic history (Figure 7.4). Therefore it is not conclusive that humans are causing global climate change.

Global dimming

Revised ☐

Global dimming refers to a reduction in global temperatures as a result of pollution.

In the 3 days after the 9/11 attacks on the World Trade Center, daily temperatures increased by an average of 1.1°C. The cause of this increase was the grounding of airlines in the interests of national security. The absence of vapour trails left behind by high-flying aircraft (Figure 7.11), and the absence of small droplets (aerosols) they form, caused this effect.

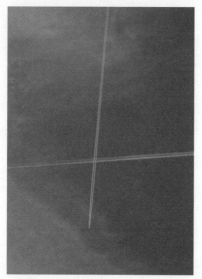

Figure 7.11 Contrails – the vapour trails that form behind aircraft

Aerosols are highly reflective and reflect solar energy, thereby blocking it from entering the lower atmosphere, which has a cooling effect. Air pollution has a similar impact. Scientists who discovered the phenomenon called it 'global dimming'.

It is possible that global dimming has slowed down global warming. Ironically, therefore, by cleaning up air pollution, climate change may be accelerated.

Scientists showed that from the 1950s to the early 1990s, the level of solar energy reaching the Earth's surface dropped 9% in Antarctica, 10% in the USA, and almost 30% in Russia due to high levels of pollution at that time.

Natural particles in clean air provide points of attachment for water. Polluted air contains far more particles than clean air (e.g. ash, soot, sulfur dioxide etc.) and therefore provides many sites for water to bind to. These tend to be smaller than natural droplets. The many small water droplets reflect more sunlight than a few larger ones, so polluted clouds reflect far more light back into space, preventing the Sun's heat from getting through.

The mechanism of global dimming works because not only are the particles over polluted areas themselves reflecting more sunlight, but also the water droplets formed around these particles reflect more light.

7.3 Climate change – mitigation and adaptation

Revised

SIGNIFICANT IDEAS
- Mitigation attempts to reduce the causes of climate change.
- Adaptation attempts to manage the impacts of climate change.

Mitigation strategies

Revised

Mitigation involves reduction and/or stabilisation of greenhouse gas emissions and their removal from the atmosphere.

Mitigation strategies to reduce greenhouse gases in general may include:

- reduction of energy consumption, for example through public transport and energy conservation measures
- reduction of emissions of oxides of nitrogen and methane from agriculture, for example using less chemical fertilisers and consuming more vegetables
- using alternatives to fossil fuels, for example more HEP, wind- and solar-energy
- geo-engineering, for example CO_2 capture from air, cloud seeding, space-based reflectors, land-based reflectors and ocean fertilisation.

Mitigation strategies to reduce carbon emissions include:

- introducing carbon taxes to make the use of fossil fuels more expensive relative to other forms of energy
- carbon trading, in an attempt to manage the amount of carbon dioxide released by different sectors/industries; this places a limit on total trading
- carbon offset schemes, in an attempt to reduce the overall impact of carbon emissions by investing in projects that cut emissions elsewhere

Mitigation strategies for carbon dioxide removal (CDR) include:

- conserving, protecting and enhancing carbon sinks through land management, for example through the UN Initiative on Reducing Emissions from Deforestation and Forest Degradation in Developing Countries (UN-REDD)
- using biomass as a fuel source, although this may take up valuable land for farming and drive up food prices
- using carbon capture and storage (CCS) either at the plant where it is produced and then storing it in a geological deposit underground, or removing it from the atmosphere with chemical sorbents, a process known as air capture. However, CSS technologies are costly and unproven.
- enhancing carbon dioxide absorption by the oceans through either fertilising oceans with compounds of nitrogen, phosphorus and iron to encourage the biological pump, or increasing upwellings to release nutrients to the surface. This increases marine food production, and removes carbon dioxide from the atmosphere.

Even if mitigation strategies drastically reduce future emissions of greenhouse gases, past emissions will continue to have an effect for decades to come.

Adaptation strategies

Revised

Adaptation strategies can be used to reduce adverse affects and maximise any positive effects. Examples of adaptations include flood defences, vaccination programmes, desalinisation plants and planting of crops in previously unsuitable climates. Adaptive capacity varies from place to place and can be dependent on financial and technological resources. Rich countries can provide economic and technological support to low-income countries.

International agreements

Revised

There are international efforts and conferences to address mitigation and adaptation strategies for climate change – for example, the Intergovernmental Panel on Climate Change (IPCC), National Adaptation Programmes of Action (NAPAs) and the United Nations Framework Convention on Climate Change (UNFCCC). The IPCC believes that carbon capture and storage are extremely important. NAPAs are a list of ranked priority adaptation activities and projects. They focus on urgent and immediate needs.

The Kyoto Protocol

The **Kyoto Protocol** (1997) gave all MEDCs legally binding targets for cuts in emissions from the 1990 level by 2008–12. The EU agreed to cut emissions by 8%, Japan by 7% and the USA by 6%. Some countries found it easier to make cuts than others.

Since 1992, when negotiations for the Kyoto Protocol first began, greenhouse gas emissions have risen by 50%. In 2010, despite a global recession, they rose by 5%.

There are three main ways for countries to keep to the Kyoto target without cutting domestic emissions:

- Install clean technology in other countries and claim carbon credits for themselves.
- Buy carbon credits from countries such as Russia where traditional heavy industries have declined and the national carbon limits are underused.
- Plant forests to absorb carbon or change agricultural practices (e.g. keep fewer cattle).

Even if greenhouse gas production is cut by between 60% and 80%, there is still enough greenhouse gas in the atmosphere to raise temperatures by 5°C.

The Kyoto Protocol deadline was the end of 2012 (Figure 7.11). However, it was only meant to be the beginning of a long-term process, not the end of one.

Paris Conference, 2015

The 2015 UN Climate Change Conference was held in Paris. Paris was taken as an example of a HIC that had decarbonised its energy production – France generates over 90% of its energy from nuclear power, hydroelectric power and wind energy. The conference resulted in the Paris Agreement on the reduction of climate change. 174 countries signed the agreement. The key objective is to limit global warming to 2°C compared with pre-industrial levels. It also seeks for zero net anthropogenic greenhouse gas emissions between 2050 and 2100. To achieve a 1.5°C goal would require zero net emissions by 2030–50.

Unlike the Kyoto Protocol, there are no country-specific goals or a detailed timetable for achieving the goals. Countries are expected to reduce their carbon usage 'as soon as possible'. However, there is no mechanism to force a country to set a specific target, nor is there any measure to penalise countries if their targets are not made.

Expert tip

You should consider the following pollution management strategies when answering exam questions:

- global – intergovernmental and international agreements (for example, the Kyoto Agreement and subsequent updates), carbon tax and carbon trading, alternative energy sources
- local – explore your lifestyle with respect to greenhouse gas emissions.

Consider both preventive and reactive strategies.

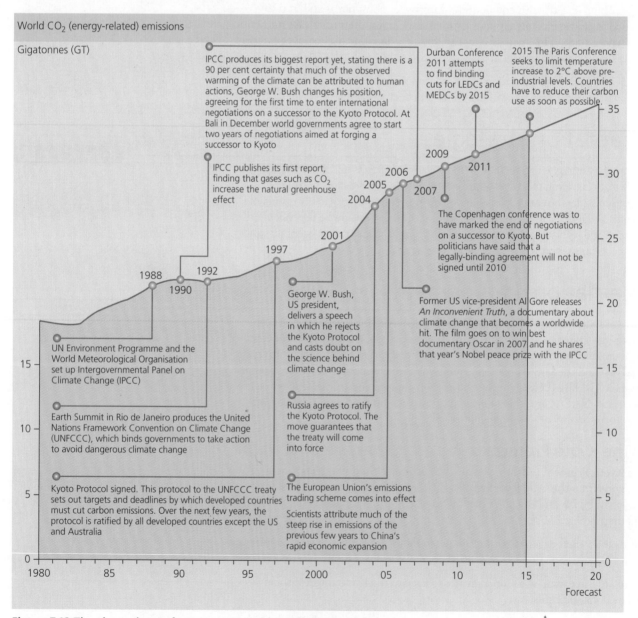

Figure 7.12 The chronology of attempts to tackle global warming

Individuals' reductions in greenhouse gas emissions

There are many ways in which individuals can reduce their own emissions of greenhouse gases:

- Use public transport.
- Walk or ride a bike for local trips.
- Turn off lights when leaving a room.
- Reduce the use of heating and air conditioning.
- Eat locally produced food.

■ **QUICK CHECK QUESTIONS**

19 In what year did the Kyoto Protocol 'run out'?
20 Identify the conference that attempted to find legally binding cuts in CO_2 emissions by 2015.

Common mistake

Some students consider preventive and reactive strategies to be the same – this is not the case.

- Preventive strategies prevent global warming from taking place. Any strategy that prevents fossil fuels from being burned is a preventive strategy.
- Reactive strategies try to treat the symptoms of global warming – attempts to reduce carbon emissions are a reaction to the build-up of greenhouse gases.

Expert tip

You should evaluate these strategies with regard to their effectiveness and the implications for MEDCs and LEDCs of reducing CO_2 emissions in terms of economic growth and national development.

Expert tip

You should be able to:

- **discuss** mitigation and adaptation strategies to deal with impacts of climate change
- **evaluate** the effectiveness of international climate change talks.

■ **QUICK CHECK QUESTIONS**

21 Distinguish between mitigation and adaptation.
22 Outline the range of geo-engineering techniques.

EXAM PRACTICE

3 Describe ecocentric and technocentric responses to global warming and justify which may be more effective in reducing the impacts of global warming. [7]

4 Justify the argument that changes need to be made in individual lifestyles to effectively address the issues of global warming. [6]

8.1 Human population dynamics

> **SIGNIFICANT IDEAS**
> - A variety of models and indicators are employed to quantify human population dynamics.
> - Human population growth rates are impacted by a complex range of changing factors.

Demographic indicators

> **Keyword definitions**
>
> **Crude birth rate (CBR)** – The number of live births per 1000 people in a population per year.
>
> **Crude death rate (CDR)** – The number of deaths per 1000 people in a population per year.
>
> **Infant mortality rate (IMR)** – The number of deaths of children less than 1 year old per 1000 live births per year.
>
> **Total fertility rate (TFR)** – The average number of births per woman of childbearing age.
>
> **Natural increase (annual growth rate)** – Found by subtracting the crude death rate (‰ – per thousand) from the crude birth rate (‰) and is then expressed as a percentage (%):
>
> $$\text{natural increase (\%)} = \frac{\text{birth rate per 1000} - \text{death rate per 1000}}{10}$$
>
> Highest growth rates are found in Africa, while lowest growth rates are in North America and Europe.
>
> **Doubling time** – The length of time it takes for a population to double in size, assuming its natural growth rate remains constant. Approximate values for it can be obtained by using the formula:
>
> $$\text{doubling time (years)} = \frac{70}{\text{growth rate (\%)}}$$
>
> **Life expectancy (E0)** – The average number of years that a person can be expected to live, usually from birth, if demographic factors remain unchanged.

Table 8.1 Selected population indicators

	UK	China	Brazil	Ethiopia
CBR (‰)	12.17	12.49	14.46	37.27
TFR	1.89	1.60	1.77	5.15
CDR (‰)	9.35	7.53	6.58	8.19
IMR (‰)	4.38	12.44	18.6	53.4
Life expectancy (years)	80.54	75.4	73.5	61.5
Growth rate (%)	0.54	0.45	0.77	2.89
Doubling time (years)	129	156	91	24

Global population change

- The world's population has grown rapidly or **exponentially**. Most of this growth is quite recent and much of it has been in south and east Asia.
- Growth is likely to take place until at least 2050. Global population doubled between 1650 and 1850, 1850 and 1920, and 1920 and 1970. It is thus taking less time for the population to double.
- Up to 95% of population growth is taking place in LEDCs. However, the world's population is expected to stabilise at about 12 billion by around 2050–80.

Figure 8.1 shows that in most regions population change increased between 1950 and 1990. The exceptions were North America and Europe. Population growth will continue to occur but should slow down by 2050.

Population growth and resources

Population growth can lead to:

- great pressures on governments to provide for their people
- increased pressure on the environment
- increased risk of famine and malnutrition
- greater differences between the richer countries and the poorer countries.

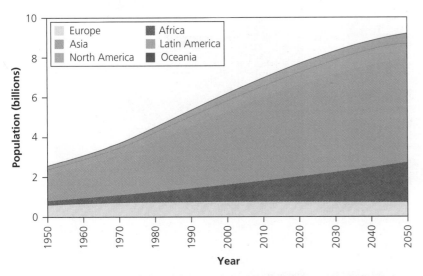

Figure 8.1 Exponential growth of the world's population, 1950–2050

> **Keyword definition**
>
> **Exponential growth** – An increasing or accelerating rate of growth, sometimes referred to as a J-shaped or J-population curve.

> **Expert tip**
>
> If you can refer to particular dates and population sizes – for example, the world's population reached 7 billion in 2011 – that will gain you extra credit.

> **Common mistake**
>
> The world's population is not currently undergoing exponential growth – it is slowing down – more of an S-population curve than a J-population curve.

■ QUICK CHECK QUESTIONS

1 **a** Approximately what was the world population in 2000?

   b Approximately what is the projected world population for 2050?

2 From the data provided in the table below, state the natural increase and the doubling time for Afghanistan and the USA.

	Afghanistan	USA
Birth rate (‰)	38.57	12.49
Death rate (‰)	13.89	8.15
Natural increase (%)		
Doubling time (years)		

Age–sex (population) pyramids

Population structure or composition refers to any *measurable* characteristic of the population. This includes the age, sex, ethnicity, language, religion and occupation of the population. Population pyramids tell us a lot of information about the age and sex structure of a population:

- A wide base suggests a high birth rate.
- A narrowing base indicates a falling birth rate.
- Straight or near vertical sides show a low death rate.
- A concave slope suggests a high death rate.
- Bulges in the slope indicate high rates of in-migration (for instance, excess elderly, usually female, will indicate retirement resorts; excess males aged 20–35 years will be economic migrants looking for work).
- Deficits in the slope show out-migration or age-specific or sex-specific deaths (war, epidemics).

Population pyramids (Figure 8.2 and 8.3) are important because they tell us about population growth. They help planners to find out how many services and facilities, such as schools and hospitals, will be needed in the future.

> **Keyword definition**
>
> **Age–sex pyramid** – A bar graph that shows the relative or absolute amount of people in a population at different ages and of different sexes.

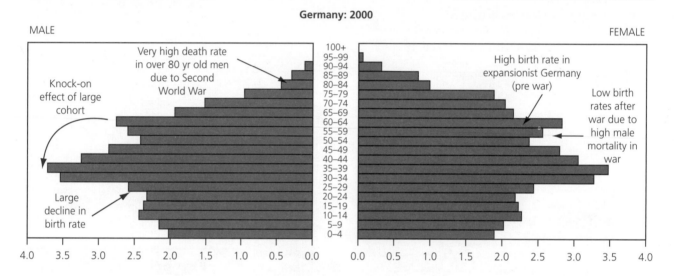

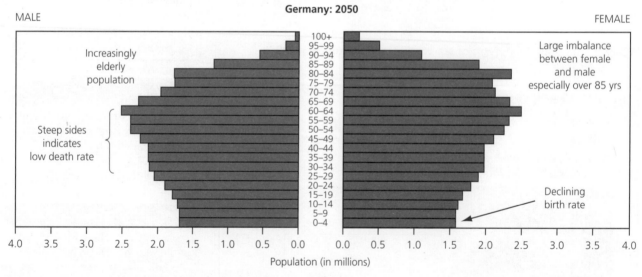

Figure 8.2 Population pyramids for Germany, 2000 and 2050

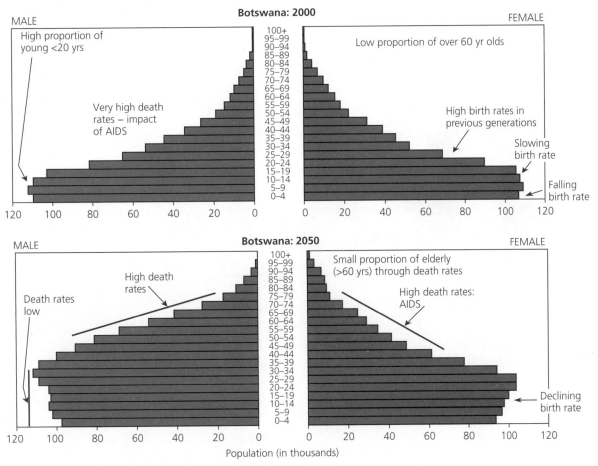

Figure 8.3 Population pyramid for Botswana, 2000 and 2050

The demographic transition model

Revised ☐

The **demographic transition model** (DTM) describes how birth rates and death rates change over time. It was developed from a study of changes in birth rates and death rates in England and Wales, and in Sweden.

The DTM is usually divided into four stages and, increasingly, a fifth stage (Figure 8.4). It is a useful diagram as it displays visually the complex patterns of change in birth rate, death rate, natural increase (i.e. the increase brought about when birth rates exceed death rates), natural decrease (the decrease in population when death rates exceed birth rates) and population growth rates.

Typically, the model shows a change from pre-industrial time, with high birth rates and death rates, to a modern society with low birth rates and death rates, and eventually an ageing population with higher death rates than birth rates.

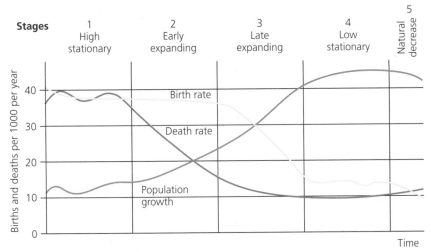

Figure 8.4 Typically five stages are identified in the demographic transition model

> **Keyword definition**
>
> **Demographic transition model** – A model that shows the change in a population from one that has high birth rates and high death rates to a country that has low birth rates and low death rates.

■ **QUICK CHECK QUESTION**

3 Describe the main characteristics of the age–sex pyramid shown below.

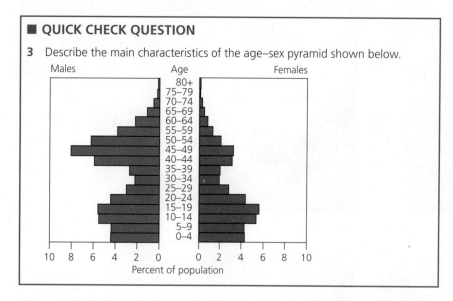

Males ⟶ Age ⟵ Females

Percent of population

Factors affecting population dynamics

Revised

■ Factors affecting birth rates and fertility rates

Changes in birth rates and fertility rates are caused by a combination of **socio-cultural**, **historical**, **religious**, **political** and **economic** factors:

- In countries where the **status of women** is low and few women are educated or have paid employment, birth rates are high.
- In countries, such as Singapore, where the status of women has improved, the birth rate has fallen.
- In general, the higher the level of parental **education**, the fewer the children. The high cost of children in a wealthy society helps to explain the falling birth rates in rich countries.
- The role of **religion** in relation to fertility rates is commonly confused. In general, most religions are **pro-natalist**, i.e. they favour large families.
- **Economic prosperity** favours an increase in the birth rate, while increasing costs lead to a decline in the birth rate.
- High infant mortality rates increase the pressure on women to have more children to offset the high mortality losses. Such births are termed **replacement births** or **compensatory births**.
- In some agricultural societies, parents have larger families to provide labour for the farm and as security for the parents in old age. Mechanisation of agriculture reduces the need for a high birth rate.

Expert tip

When describing a map of global variations in birth rates or fertility rates, look out for the highest, lowest, trends and exceptions. Also, make sure you use the key, and give some examples of each.

Common mistake

Crude birth rates and crude death rates are called crude because they do not take into account the age structure of a population.

■ Patterns of mortality

- At the global scale, the pattern of mortality in rich countries differs from that in poor countries.
- In rich countries, as a result of better **nutrition**, **health care** and **environmental conditions (housing, safe water, proper sanitation)**, the death rate has fallen steadily, with very high life expectancies (75+ years).
- In many of the very poor countries, high death rates and low life expectancies are still common, although both have shown steady improvement over the past few decades. This trend, unfortunately, has been reversed as a consequence of AIDS in some parts of the world, especially sub-Saharan Africa and Russia.
- Some populations, such as those in retirement towns and especially in the older industrialised countries, have very high life expectancies and this in turn results in a rise in the CDR. Countries with a large proportion of young people will have much lower death rates.

National and international policies affecting population dynamics

- Many policy factors influence human population growth. **Pronatalist** policies refer to attempts to increase the population, for example by providing maternity care, accessible and affordable child-care and a child-allowance for every child. **Anti-natalist** policies are those that try to limit the number of children born either through increased use of contraception, forced abortions or sterilisations. Other policies can indirectly affect the death rate and the birth rate.
- Agricultural development, improved public health and sanitation, and better service infrastructure can stimulate rapid population growth by lowering mortality without significantly affecting fertility.
- Some parents may want large families as they may depend on their children during their old age.
- Following urbanisation, birth rates often fall, as more women have jobs in the **formal** and **informal sectors**.
- Increased awareness about family planning can increase its usage and help reduce the birth rate.
- Cultural or religious influence on contraception usage/non-usage can decrease/increase fertility.
- Boys being more valued than girls in some cultures can increase fertility so that more boys are born.
- Policies that target female education and female participation in the job market are believed to be the most effective method for reducing population pressure.
- Policies can encourage immigration to facilitate gaps in the labour market in countries with falling birth rates. As most economic migrants are young, they will increase the proportion of people in a country of child-bearing age and could lead to an increase in the birth rate.
- A country can also affect birth and death rates through its economic policies. An older retirement age could lead to an increase in death rates among the elderly and a falling life expectancy. Countries that are forced to re-structure their economy in return for aid may reduce the amount of social welfare spending, which could impact on both birth rates and death rates. Organisations such as the World Bank, International Monetary Fund (IMF) and the World Trade Organization (WTO) have a major influence on poor countries and on the economic and social policies that they must follow.

CASE STUDY

CONTRASTING POPULATION POLICIES IN CHINA AND SINGAPORE

The most famous anti-natalist policy is China's one-child policy, which was introduced in 1979. It limited the majority of Chinese families to just one child. Without it, China's population would now be 400 million larger.

Some critics believe that China's fertility would have come down regardless, as a result of urbanisation, industrialisation, improved female education and more working women. Policies directed towards the education of women, enabling women to have greater personal and economic independence, may be the most effective method for reducing population pressure. China's one-child policy lasted between 1979 and 2015, when it was replaced by a two-child policy.

Singapore also followed an anti-natalist policy, but changed during the 1980s to a pro-natalist policy. Its fertility rate had dropped to below 1.25, and the workforce was decreasing in size. The government offered incentives to families to have three or more children if they could afford them.

Despite the incentives, Singapore's fertility rate has remained low, as women continue to play an active role in the workforce, and are choosing jobs ahead of having children.

Common mistake

Some students state that the one-child policy is the reason for the fall in the birth rate in China. It is just one factor among many – female education, female participation in the workforce, rising material ambitions and industrialisation are also important.

Expert tip

Even if the birth rate falls, there could be an increase in population size. This is known as **population momentum**. This is because of a large proportion of the population entering the reproductive years and is common in countries with a youthful population.

Expert tip

You should be able to:

- **calculate** values of CBR, CDR, TFR, DT and NIR
- **explain** the relative values of CBR, CDR, TFR, DT and NIR
- **analyse** age–sex pyramids and diagrams showing demographic transition models
- **discuss** the use of models in predicting the growth of human populations
- **explain** the nature and implications of growth in human populations
- **analyse** the impact that national and international development policies can have on human population dynamics and growth
- **discuss** the cultural, historical, religious, social, political and economic factors that influence human population dynamics.

■ **QUICK CHECK QUESTIONS**

4 Briefly explain the meaning of the terms *anti-natalist* and *pro-natalist*.

5 What is believed to be the most effective way to reduce the birth rate?

EXAM PRACTICE

1 Describe the population growth in Europe, as shown in Figure 8.1. [3]

2 Explain the implications of exponential growth in human populations. [4]

3 Examine how socio-cultural and economic factors affect birth rates. [8]

4 Analyse the demographic transition model as a way of showing population change. [6]

5 a Study the population pyramid for Haiti in 2010. Describe, giving a reason, how Haiti's population is likely to change over the next 20 years. [2]

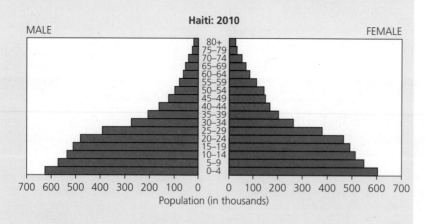

Haiti: 2010

b Suggest reasons why Haiti's population might not expand rapidly over the next 20 years. [3]

8.2 Resource use in society

Revised

SIGNIFICANT IDEAS
- The renewability of natural capital has implications for its sustainable use.
- The status and economic value of natural capital is dynamic.

Renewable and non-renewable natural capital

Revised ☐

Renewable natural capital, such as living species and ecosystems, is self-producing and self-maintaining and uses solar energy and photosynthesis. This natural capital can yield marketable goods such as wood, but may also provide unaccounted essential services when left in place, for example, climate regulation. Renewable natural capital also includes non-living features, such as groundwater and the ozone layer, that are dependent on the solar 'engine' for renewal.

Non-renewable natural capital (except on a geological timescale), such as fossil fuels, minerals and soils, are resources in which use implies partial depletion of the stock.

Common mistake

Some students think that nuclear energy is a renewable form of energy – it is non-renewable. It is, however, considered to be an alternative form of energy (alternative to the carbon-rich fuels such as oil, natural gas and coal). Politically, some countries have considered nuclear energy alongside renewables, such as HEP and wind energy. However, uranium will run out just as fossil fuels will.

Expert tip

The terms *natural resources* and *natural capital* are often substituted for each other. However, the term 'natural resources' suggests that resources are there to be used, whereas natural capital suggests something to be managed to produce an income or a return on an investment.

■ QUICK CHECK QUESTIONS

6   Define the terms *renewable natural capital* and *non-renewable natural capital*.

7   For the following resources, identify whether each is considered to be renewable or non-renewable: groundwater; potatoes; iron ore in rocks; sheep's wool; the ozone layer; water used for the generation of HEP.

■ Renewable natural capital and sustainability

■   The term **sustainability** refers to the use of global resources at a rate that allows natural regeneration and minimises damage to the environment.

■   Any society that supports itself in part by depleting natural capital is unsustainable. If human wellbeing is dependent on the goods and services provided by certain forms of natural capital, then long-term use rates should not exceed rates of natural capital renewal.

■   Sustainability means living within the means of nature, on the 'interest' or sustainable income generated by natural capital – for example, harvesting renewable resources at a rate that will be replaced by natural growth demonstrates sustainability.

■   Sustainability focuses on the rate of resource use and suggests maintaining a balance between resource use and natural income.

Keyword definition

Sustainability – The use of global resources at a rate that allows natural regeneration and minimises damage to the environment.

Expert tip

Deforestation can be used to illustrate the concept of sustainability and unsustainability.

●   If the rate of forest removal is less than the annual growth of the forest, then the forest removal is sustainable.

●   If the rate of forest removal is greater than the annual growth of the forest, then the forest removal is unsustainable.

Common mistake

Some students think that nuclear energy is a renewable form of energy – it is non-renewable. It is, however, considered to be an alternative form of energy (alternative to the carbon-rich fuels such as oil, natural gas and coal). Politically, some countries have considered nuclear energy alongside renewables, such as HEP and wind energy. However, uranium will run out just as fossil fuels will.

Natural capital – goods and services

Natural capital can be assessed in different ways (Figure 8.5).

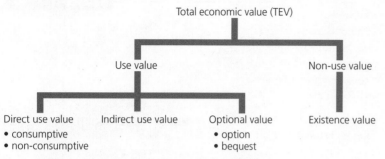

Figure 8.5 Methods of assessing natural capital

> **Keyword definitions**
>
> **Natural capital** – Natural resources that are managed to provide goods and services for societies.
>
> **Natural income** – The portion of natural capital (resources) that is produced as 'interest', i.e. the sustainable income produced by natural capital.

- ▪ **Direct use values** are ecosystem goods and services that are directly used by humans – most often by people visiting or residing in the ecosystem.
- ▪ **Consumptive use values** include harvesting food products, timber for fuel or housing, medicinal products and hunting animals for food and clothing.
- ▪ **Non-consumptive use values** include recreational and cultural activities that do not require harvesting of products.
- ▪ **Indirect use values** are derived from ecosystem services that provide benefits outside the ecosystem itself (e.g. natural water filtration, which may benefit people downstream).
- ▪ **Optional values** are derived from the potential future use of ecosystem goods and services not currently used – either by yourself (**option value**) or your future offspring (**bequest value**).
- ▪ **Non-use values** include aesthetic and intrinsic values, and are sometimes called **existence values**. They have no market price.

Ecosystems that are valued on aesthetic or intrinsic grounds may not provide identifiable goods or services, and so remain unpriced or undervalued from an economic viewpoint.

There are many examples of places or ecosystems that have an important national identity – for example, Mount Fuji in Japan or Mount Kilimanjaro in Tanzania. Uluru (Ayers Rock) in Australia has great spiritual value for the Aboriginal population. Such areas or ecosystems have intrinsic value from an ethical, spiritual or philosophical perspective, and are valued regardless of their potential use to humans.

There are many attempts to value nature – for example, biodiversity and rate of depletion of natural resources – so that they can be weighed more rigorously against more common economic values (for example, gross national income (GNI)). However, to a large extent these valuations are impossible to quantify realistically. Not surprisingly, much of the sustainability debate centres on the problem of how to weigh conflicting values in our treatment of natural capital.

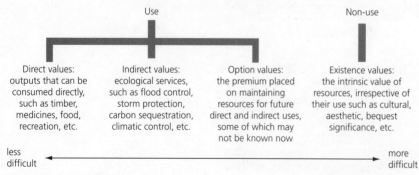

Figure 8.6 Level of difficulty in assessing the economic value of natural capital

> **Common mistake**
>
> Direct and indirect values can be quantified, up to a point. However, option values and non-use values are nearly impossible to quantify – but should not be ignored.

Resources and natural income

- Resources are everything that is useful to mankind.
- They include air, water, soil, people, education, fossil fuels, ecosystems and so on.
- People can get many benefits from resources.
- Some environmentalists describe resources as 'natural capital'.
- They often make a comparison with saving money (capital) in a bank. At the end of the year, the savings (capital) may have gained some interest.
- Likewise, with natural resources (**natural capital**), over time these may produce more resources, and so people can live off the 'interest' – this is known as '**natural income**'.
- Renewable resources can produce natural income indefinitely in the form of valuable goods and services.
- These goods and services include marketable goods such as timber and food or may be in the form of ecological services such as the flood and erosion protection provided by forests (services) and climate regulation.
- Some of these services are impossible to quantify.
- Non-renewable resources, such as oil and coal, generate wealth but can be used only once in a human lifetime.

Ecosystem services

Examples of ecosystem services are listed in Table 1.5 on page 16. There are four main types of ecosystem service:

- **Supporting services** are the essentials for life and include primary productivity, soil formation and the cycling of nutrients.
- **Regulating services** are a diverse set of services and include pollination, regulation of pests and diseases and production of goods such as food, fibres and wood. Other services include climate and hazard regulation and water quality regulation.
- **Provisioning services** are the services people obtain from ecosystems such as food, fibre, fuel (peat, wood and non-woody biomass) and water from aquifers, rivers and lakes. Goods can be from heavily managed ecosystems (intensive farms and fish farms) or from semi-natural ones (such as by hunting and fishing).
- **Cultural services** are derived from places where people's interaction with nature provides cultural goods and benefits. Open spaces – such as gardens, parks, rivers, forests, lakes, the sea-shore and wilderness areas – provide opportunity for outdoor recreation, learning, spiritual wellbeing and improvements to human health.

> **Expert tip**
>
> The terms *natural resources* and *natural capital* are often substituted for each other. However, 'natural resources' suggests that resources are there to be used, whereas 'natural capital' suggests something to be managed to produce an income or a return on the investment.

> **Expert tip**
>
> How can we quantify values such as aesthetic value, which are inherently qualitative? Have an example ready – such as Mount Fuji or Uluru – to support any statements that you make about the intrinsic value of nature in an examination.

> ## Keyword definitions
>
> **Economic value** – The market price of the goods and services a resource produces.
>
> **Ecological value** – Resources with no formal market price: soil erosion control, nitrogen fixation and photosynthesis are all essential for human existence but have no direct monetary value, although some estimates have been made.

> **Expert tip**
>
> The supporting services, including primary production and nutrient cycling, are not listed for the individual habitats as they are considered necessary for the production of all other ecosystem services.

> ## ■ QUICK CHECK QUESTIONS
>
> 8 Identify *two* forms of natural income that might be derived from the damming of rivers.
>
> 9 Suggest another way of valuing the temperate deciduous woodland ecosystem other than for its economic value.

The dynamic nature of natural capital

Revised

Resources change over time – they are affected by cultural, economic and technological factors. For example, in the Middle Ages oil was used to treat wounds, whereas in the twentieth and twenty-first centuries it has been used as a fuel and as an industrial raw material for the plastics and chemical fertiliser industry. Uranium only became a valuable resource in the mid-twentieth century because of the development of nuclear technology.

■ Shale gas

Shale rocks have recently become an important potential resource because of the shale gas (natural gas) that they contain, and many countries are trying to develop their shale gas resources.

■ In the USA, shale gas has 'taken off' because of a plentiful supply of shale rock, the available technology (pipelines that had been used to carry natural gas and oil are now carrying shale gas) and the country's determination to achieve 'energy security'.

■ In contrast, China, although it has large deposits of shale rocks in areas such as Sichuan province, is less likely to develop shale gas as there is a risk of triggering earthquakes, such as the one that devastated Sichuan in 2008, killing over 66 000 people.

■ In the UK, there are differing reports regarding the amount of shale rock deposits and the location of the rock with best potential for shale gas. Deposits in the south of the country are less likely to be developed because of a more politicised middle-class population, whereas in the north of the country, the need for employment and investment could prove more powerful than environmental considerations.

Common mistake

It is quite common to think of energy resources when thinking of the dynamic nature of resources. However, other resources – such as ecosystems – can change in importance over time.

Expert tip

You should be able to:

● **outline** an example of how renewable and non-renewable natural capital has been mismanaged

● **explain** the dynamic nature of the concept of natural capital.

8.3 Solid domestic waste

`Revised ▢`

> ### SIGNIFICANT IDEAS
>
> ● Solid domestic waste (SDW) is increasing as a result of growing human populations and consumption.
> ● Both the production and management of SDW can have significant influence on sustainability.

Types of solid domestic waste

`Revised ▢`

There are many types of solid domestic (or municipal) waste (rubbish or garbage). In an HIC these generally consist of:

■ paper/packaging/cardboard (20–30%)
■ glass (5–10%)
■ metal (less than 5%)
■ plastics (5–10%)
■ organic waste from kitchen or garden, including waste wood (20–50%)
■ textiles (less than 5%)
■ nappies (diapers) (2%)
■ electrical appliances such as computers/fridges (known as WEEE – waste electrical and electronic equipment – see Figure 8.7) (less than 5%)
■ rubble/bricks (less than 1%)
■ ash (less than 1%).

■ QUICK CHECK QUESTIONS

10 Explain *three* ways in which the status of large herbivores, such as red deer, as a natural resource has changed over time.

11 Suggest how a woodland ecosystem as a resource might change over time.

Expert tip

Keep up to date with major developments and news items. For example, the nuclear energy industry in a number of countries, including Japan and Germany, suffered a major setback following the tsunami and explosion at the Fukushima Daiichi nuclear power station in March 2011.

The amount will vary from place to place (Figure 8.7), and over time. The total volume of waste generated can be over 800 kg per person per year.

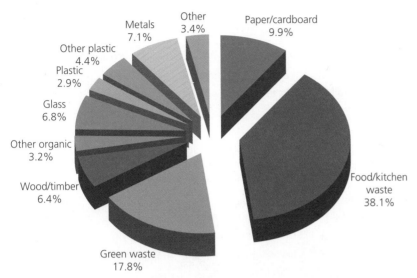

Figure **8.8** Composition of household waste, Victoria, Australia

Figure 8.7 The WEEE man at the Eden Project, Cornwall, UK, by sculptor Paul Bonomini, standing 7 m tall and weighing 3.3 tonnes

Non-biodegradable waste

Revised ☐

The abundance and prevalence of non-biodegradable pollution, such as plastics, batteries and e-waste, has become a major environmental issue. China receives about 70% of the world's electronic waste. Processing and recycling electronic waste has come at a high cost for many living in and around the factories. In Guiyu, the so-called e-waste capital of the world, workers suffer from lead poisoning, cancer and elevated risk of miscarriage. Soil, water and air have all become polluted.

> **Keyword definition**
>
> **Biodegradable** – Capable of being broken down by natural biological processes – for example, through the activities of decomposer organisms.

> **Expert tip**
>
> You should consider your own and your community's generation of waste. Consider the different types of material – for example, paper, glass, metal, plastics, organic waste (kitchen or garden), packaging – as well as their total volume. Keep a record of your waste for a week.

■ **QUICK CHECK QUESTIONS**

12 Identify the type of waste that we might expect to increase around Christmas time.

13 Explain how the volume of waste is likely to vary for a household in Tokyo (Japan) and one in Kabul (Afghanistan).

Common mistake

Nappies are a form of solid domestic waste, but faeces are not.

Waste disposal methods

Revised ☐

■ Recycling

Table **8.2** Advantages and disadvantages of recycling

Advantages	Disadvantages
● Reduced amount of energy required to recycle compared with exploiting the resource	● It involves transport of sometimes heavy, bulky goods, so requires lots of energy
● It reduces the amount of resources used	● It can produce toxic waste
● It maintains stocks of non-renewable and replenishable resources	● It can be labour intensive
● It reduces the amount of material in landfill sites	● It can be uneconomic in terms of demand and supply factors
● It can be used to make new products	
● It reduces greenhouse gas emissions	
● It creates new job opportunities	

Figure 8.9 Recycling

◼ Reuse

Table 8.3 Advantages and disadvantages of reuse

Advantages	Disadvantages
<ul><li>Little energy is used</li><li>It provides cheap resources for people of limited means</li></ul>	<ul><li>It can require energy to clean the products being reused (e.g. milk bottles)</li><li>The products may be heavy to transport (e.g. milk bottles)</li><li>The products will eventually wear out and must be disposed of</li></ul>

◼ Composting

Table 8.4 Advantages and disadvantages of composting

Advantages	Disadvantages
<ul><li>It produces fertiliser</li><li>It reduces the volume of waste</li><li>It reduces the use of chemical fertilisers</li></ul>	<ul><li>It produces unpleasant smells</li><li>It can attract vermin if not done properly</li><li>It requires effort and space</li><li>It takes time</li></ul>

◼ Incineration

Table 8.5 Advantages and disadvantages of incineration

Advantages	Disadvantages
<ul><li>It reduces the volume of waste, thereby reducing the need for landfill</li><li>The heat produced can be used in place of burning fossil fuels</li><li>It kills pathogens</li><li>It produces ash for construction</li><li>It is a way of producing energy from waste</li></ul>	<ul><li>It releases toxic chemicals</li><li>It produces greenhouse gases</li><li>Ash still needs disposal</li><li>It is expensive</li><li>There may be considerable community resistance to the building of new incinerators</li></ul>

◼ Landfill

Table 8.6 Advantages and disadvantages of landfill

Advantages	Disadvantages
<ul><li>It is a cheap and easy way to dispose of waste</li><li>It is a way of producing energy (in the form of methane) from waste</li><li>Relatively limited amounts of time and labour are required</li><li>It can create land (e.g. in Hong Kong)</li></ul>	<ul><li>Leachate can pollute watercourses and groundwater.</li><li>It gives off unpleasant odours</li><li>It increases vermin, which can cause disease to spread</li><li>It produces methane, which is a greenhouse gas</li><li>It takes up land area, and in many places (e.g. New York, USA) there is a limited amount of space available for landfill</li><li>There is potential of subsidence and/or contamination of future building land</li></ul>

Pollution management strategies

Revised ◻

There are a number of strategies that can be used to manage solid domestic waste (Figure 8.10):

- ◼ Altering human activity – for example, by reduced consumption, increased recycling, reuse of materials and composting of food waste.
- ◼ Controlling the release of pollutants – standards can be set and targets introduced for increased rates of recycling of waste, reuse of materials, taxes or charges on plastic bags, and increased charges for waste collection.
- ◼ Cleaning and restoring environments by removing pollutants from the environment – for example, use of energy for waste (EfW) schemes/waste to

energy programmes, reclaiming landfill sites and removing pollutants such as plastics from the environment (e.g. from the Great Pacific Garbage Patch).

Process of pollution　　　**Strategies for reducing impacts**

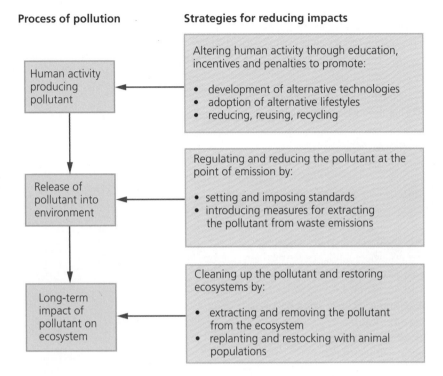

Figure 8.10 Approaches to pollution management

These strategies are influenced by a variety of factors.

■ Factors affecting the choice of waste disposal

At a national scale there are a number of factors that affect the choice of waste disposal. These include:

- ■ government policy (e.g. strategy to encourage recycling)
- ■ population density and the amount of land available for landfill
- ■ involvement in international agreements to cut greenhouse gases or dumping at sea
- ■ involvement of significant environmental pressure groups (e.g. Greenpeace) in influencing attitudes
- ■ geographic/climatic characteristics (e.g. access to coastline)
- ■ economic considerations (e.g. costs of energy and transport).

> **Expert tip**
>
> You should be able to:
>
> • **evaluate** SDW disposal options
> • **compare** and **contrast** pollution management strategies for SDW
> • **evaluate**, with reference to Figure 8.10, pollution management strategies for SDW by considering recycling, incineration, composting and landfills.

> **■ QUICK CHECK QUESTIONS**
>
> **14** State *two* disadvantages of recycling.
> **15** State *two* advantages of landfill.

> **Expert tip**
>
> Many poorer communities are much better at recycling and reusing materials. In Dharavi, Mumbai, for example, recycling and reuse forms the basis of many industries.

> **Common mistake**
>
> Recycling and reuse is not carbon-free. Many greenhouse gases are used in the transport and cleaning of materials.

> **EXAM PRACTICE**
>
> 6 Explain how the use of waste to generate energy can produce greenhouse gases.　　　[4]

8.4 Human population carrying capacity

Revised ☐

> **SIGNIFICANT IDEAS**
>
> • Human population carrying capacity is difficult to quantify.
> • The ecological footprint (EF) is a model that makes it possible to determine whether human populations are living within carrying capacity.

Carrying capacity

By examining carefully the requirements of a given species and the resources available, it might be possible to estimate the **carrying capacity** of that environment for the species. This is problematic in the case of human populations for a number of reasons:

- The range of resources used by humans is usually much greater than for any other species.
- Furthermore, when one resource becomes limiting, humans show great ingenuity in substituting one resource for another, for example plastic for glass, or shale gas for coal and oil.
- Resource requirements vary according to lifestyles, which differ over time and from population to population. For example, a Maasai herdsman uses far fewer resources than an urban dweller in a rich country.
- Technological developments give rise to continual changes in the resources required and available for consumption, for example the increase in nuclear power since the 1950s.
- Human populations regularly import resources from outside their immediate environment, which enables them to grow beyond the boundaries set by their local resources and increases their carrying capacity. The import of food (and even cut flowers) from countries such as Kenya and Zimbabwe in Africa into Europe is a good example. China's purchase of land in Ethiopia and Sudan is another. In this case food is grown and exported to China.
- While importing resources in this way increases the carrying capacity for the local population, it has no influence on global carrying capacity.
- All these variables make it practically impossible to make reliable estimates of carrying capacities for human populations.

Certain areas can 'carry' more people than others. For example, areas with warm, wet climates and fertile soils can support large population densities. By contrast, areas that are too hot, too dry, too cold or too wet will be unable to support many people because they cannot produce sufficient food.

However, in the globalised world, with increased trade, it is possible to get goods to cold areas or hot and dry regions, so that these areas can now support a larger resident population through the import of food and water. Examples include Dubai in the Middle East and the research stations in Antarctica.

■ Increasing carrying capacity

Human carrying capacity is determined by the rate of resource consumption, the level of pollution and also the extent of recycling, reuse and reduction in the use of resources.

The **limits to growth model** (Figure 8.11) was developed in the early 1970s. It predicted that the limits to the growth of the human population would be reached by 2100. This is sometimes called a neo-Malthusian view after Thomas Malthus, who suggested in 1798 that the growth of the human population would outstrip the ability of the Earth to provide sufficient food resources for the population. However, it also suggested that it would be possible to change these projections.

The optimistic view was championed by Esther Boserup, whose views have often been summarised by the phrase 'necessity is the mother of invention'. There are a number of ways in which food production, for example, could be increased:

- growing crops in nutrient-enriched water – hydroponics
- use of high-yielding varieties (HYVs) of plants and selective breeding of animals
- greater use of irrigation and fertilisers
- land reclamation – from the sea, draining of wetlands and terracing of steep slopes
- growing crops in greenhouses.

In terms of energy, there has been:

- use of new resources, such as shale gas
- greater development of alternative energy, such as HEP, solar and wind energy
- increased energy conservation – in the home, in public buildings, in industry and in transport.

Keyword definitions

Carrying capacity – The maximum number of a species or 'load' that can be sustainably supported by a given environment.

■ QUICK CHECK QUESTIONS

16 Define the term *carrying capacity*.

17 Explain why it is difficult to give a precise value for a country's carrying capacity for a human population.

Common mistake

Students commonly confuse carrying capacity with ecological footprint. The carrying capacity is the maximum number of a species (*people*) that can be sustainably supported by a given environment. In contrast, the ecological footprint refers to the *area of land* and water required to support a defined human population at a given standard of living.

Expert tip

Carrying capacities are not static – they can increase or decrease over time. The optimistic point of view is that carrying capacity will be increased through technological improvements (e.g. irrigation, fertilisers, GM food). Pessimists state that the Earth is a finite resource that can only sustain a certain level of population.

As awareness of the problem of resource depletion increases, measures are being taken to tackle the issue. On the other hand, as once poorer countries industrialise, for example Korea, China, India and Vietnam, and standards of living rise, there is increased demand for resources, including food, water and energy.

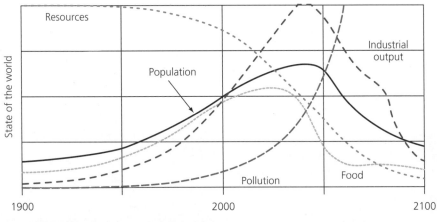

Figure 8.11 The limits to growth model

■ QUICK CHECK QUESTIONS

18 Describe the population curve shown in Figure 8.11.

19 Comment on the nature of the resources curve shown in Figure 8.11.

EXAM PRACTICE

7 'Earth provides enough to satisfy every man's need, but not every man's greed.' (Mahatma Gandhi)

'We do not inherit the Earth from our ancestors: we borrow it from our children.' (traditional Kenyan proverb and Native American saying)

With reference to the concepts of carrying capacity and environmental value systems, discuss the two quotations above. [7]

8 Justify your response to the claim that the human population has exceeded the Earth's carrying capacity. [7]

Common mistake

Some students accept the changes suggested by the limits to growth model. Many of these changes might not happen, depending on the choices that people make. With increased recycling and reuse of resources, for example, human carrying capacity can be increased.

Ecological footprints

Revised

The **ecological footprint** of a population is the area of land, in the same vicinity as the population, that would be required to provide all the population's resources and assimilate its wastes. The EF is a useful model for assessing the demands that human populations make on their environment. It consists of eight main categories, as outlined in Table 8.7.

Table 8.7 Categories contributing to the EF

Land category	Land use	Land use category
Energy land	Land 'appropriated' by fossil fuel energy use	Energy or CO_2
Consumed land	Built environment	Degraded land
Currently used land	Gardens	Reversibly built environments
	Crop land	Cultivated systems
	Pasture	Modified systems
	Managed forest	Modified systems
Land of limited availability	Untouched forest	Productive natural ecosystems
	Non-productive areas	Ice caps, deserts

Keyword definition

Ecological footprint – The area of land and water required to support a defined human population at a given standard of living. The measure takes account of the area required to provide all the resources needed by the population, and the assimilation of all wastes.

Table 8.8 Advantages and disadvantages of the EF concept

Advantages	Disadvantages
It is a useful snapshot of the sustainability of a population's lifestyle	It does not include all information on the environmental impacts of human activities
It provides a means for individuals or governments to measure their impact and to identify potential changes in lifestyle	It is only a model so it is a simplification and lacks precision
	It uses approximations of actual figures that cannot be accurately calculated
It is a popular symbol for raising awareness of environmental issues	It does not show the types of resource used – it shows only total resources
	It is negative in approach, so could be perceived as demotivating

■ Calculating ecological footprints

Although the accurate calculation of an EF might be very complex, an approximation can be calculated using per capita food consumption and per capita CO_2 emission in the formulae below:

per capita land requirement for food production (ha) =

$$\frac{\text{per capita food consumption (kg yr}^{-1})}{\text{mean food production per hectare of local arable land (kg ha}^{-1}\text{ yr}^{-1})}$$

per capita land requirement for absorbing waste CO_2 from fossil fuels (ha) =

$$\frac{\text{per capita } CO_2 \text{ emission (kg C yr}^{-1})}{\text{net carbon fixation per hectare of local vegetation (kg C ha}^{-1}\text{ yr}^{-1})}$$

The total land requirement (EF) can then be calculated as the sum of these two per capita requirements, multiplied by the total population.

This calculation clearly ignores the land or water required to: provide any aquatic and atmospheric resources; assimilate wastes other than carbon dioxide; produce the energy and material subsidies imported to arable land for increasing yields; replace loss of productive land through urbanisation; and so on. However, as a model, it is able to provide a quantitative estimate of human carrying capacity. It is, in fact, the inverse of carrying capacity as it refers to the area required to sustainably support a given population rather than the population that a given area can sustainably support.

■ QUICK CHECK QUESTIONS

20 Define the term *ecological footprint*.

21 Identify the *eight* main categories that are used to assess a society's ecological footprint.

Common mistake

National EFs hide variations in use. Some people in a society have a very large EF, whereas others have a small EF.

Expert tip

There are different approaches to the study of EFs. For example, ecocentrists suggest that the only way we can reduce our EF is by working with nature. In contrast, technocentrists believe that we can reduce our EF with the use of technology.

Worked example				
	Per capita food (grain) consumption (kg yr^{-1})	Mean food (grain) production per hectare of local arable land (kg ha^{-1} yr^{-1})	Per capita CO_2 emission (kg C yr^{-1})	Net carbon fixation per hectare (kg C ha^{-1} yr^{-1})
Brazil	210	2919	1900	10 000
Canada	710	3031	16 900	4000

Brazil's EF per person is thus:

$$\frac{210}{2919} + \frac{1900}{10\,000} = 0.072 + 0.19 = 0.26\,\text{gha}$$

where gha stands for global hectares. One global hectare is the average productivity of all biologically productive areas (in ha) in a given year.

Canada's ecological footprint per person is:

$$\frac{710}{3031} + \frac{16\,900}{4000} = 0.23 + 4.23 = 4.46\,\text{gha}$$

Note that these are very simplified footprints because they only take into account two categories. The actual ecological footprints for both countries would be much larger.

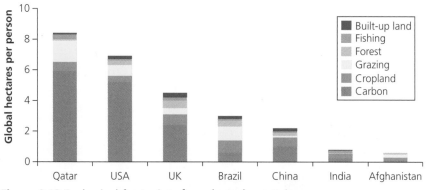

Figure 8.12 Ecological footprints for selected countries

■ **QUICK CHECK QUESTIONS**

22 State the units that ecological footprints are measured in.

23 What is the biggest contributor to the ecological footprint of:

a   Qatar

b   Brazil

c   India

as suggested in Figure 8.12.

Common mistake

Some students forget to give any units – ecological footprints refer to the size of land/water needed to support a population. The units should be given as hectares (ha) or global hectares (gha).

Expert tip

Be aware of some of the differences in the size of ecological footprint for contrasting countries, as well as differences in the structure (make-up) of the footprint, as shown in Figure 8.12.

■ Ecological footprints in LEDCs and MEDCs

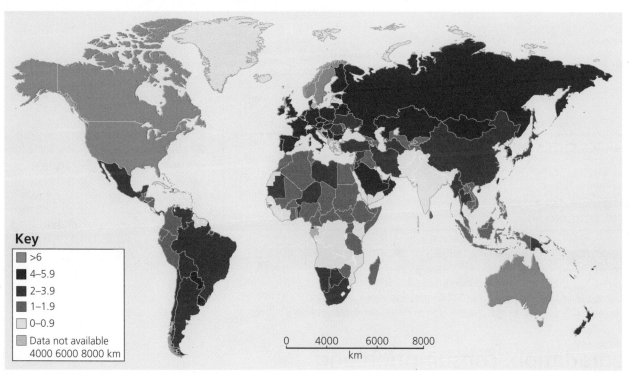

Figure 8.13 Global EFs (gha per person)

A number of factors help explain the differences in the EFs of populations in LEDCs and MEDCs:

■ EFs can be related to stages of the demographic transition model (DTM).
■ Generally there is an increase in the EF with each stage of the DTM. However, highly developed countries may now be reducing EFs through, for example, energy efficiency strategies.
■ Energy use is much greater in later stages of the DTM as people have more appliances.
■ Greater wealth leads to higher consumption of goods and natural resources such as water.

- There is more use of transport/travel by individuals in stages 4 and 5 of the DTM, so more carbon emissions occur.
- In stages 4 and 5 more goods are imported, so there are more pollutants, food miles and carbon emissions.
- There is also greater production of domestic waste in stages 4 and 5, so more area is needed to absorb it.
- Increasing levels of industrialisation from stage 3 onwards lead to more pollutants.
- Populations more dependent on fossil fuels have higher CO_2 emissions.
- People in later stages of the DTM tend to eat more meat than those in earlier stages.
- Data for food consumption are often given in grain equivalents, so that a population with a meat-rich diet would tend to consume a higher grain equivalent than a population that feeds directly on grain.
- In MEDCs, about twice as much energy in the diet is provided by animal products than in LEDCs.
- Grain production will be higher with intensive farming strategies.
- The EF for a meat eater is likely to be much larger than for a vegetarian because production of meat requires a greater energy input than growing crops.
- Meat eaters are eating at a higher trophic level, which means more land is required to create feed for livestock and for raising livestock.
- Transport of meat for processing increases the footprint size.

Table 8.9 EFs for China, India, USA and the UK

Country	Population	EF per person (gha)
China	1.3 billion	1.84
India	1.15 billion	1.06
USA	303 million	12.22
UK	60 million	6.29

■ QUICK CHECK QUESTIONS

24 Describe the main pattern of global ecological footprints as shown in Figure 8.12.

25 Using the data in Table 8.9, work out the total ecological footprint for each of the four countries.

Common mistake

When calculating EFs from data such as those in Table 8.9, do not forget to check whether you are dealing with millions (six zeros after the number) or billions (nine zeros after the number). It is easily done!

Expert tip

Different sources produce different estimates of a country's EF. For example, the data used in Figure 8.12 and Table 8.9 suggest different values for the USA's EF. However, both suggest that it is large, just as both suggest that India's is small.

Degradation, consumption and population growth

Revised ▢

As human population continues to grow, and standards of living increase, the consumption of resources increases, and this could limit population growth (see the limits to growth model on page 160). This could lead to pollution and degradation of the environment, which could limit future population growth. If people do not adopt a more sustainable lifestyle, the risk of carrying capacity being exceeded remains high, and increases the potential for a population crash.

As countries develop and improve their standard of living (material economic growth), their use of resources increases. The use of technology reinforces the consumption of resources. For example, in MEDCs and in newly industrialising countries (NICs) such as India and China, the use of cars, computers and electrical goods has increased, thereby increasing the demand for more energy resources. Many believe that the world is reaching its carrying capacity.

The world has been described as a 'globalised consumer culture', in which resources are extracted, manufactured into goods, transported, stored, sold and thrown away, to make way for new goods. The demand for consumer goods has increased dramatically in the last 30 years, putting the world's resources under great pressure.

The role of technology

Revised ☐

Because technology plays such a large role in human life, many economists argue that human carrying capacity can be expanded continuously through technological innovation. Like Boserup, the technocrats believe that 'necessity is the mother of invention', and that human flexibility and intelligence will find a solution.

If we learn to use energy and materials twice as efficiently, we can double the population or the use of energy without necessarily increasing the impact (load) imposed on the environment. However, many of the 'solutions', such as HEP and nuclear energy, require vast amounts of fossil fuels for their construction.

To compensate for foreseeable population growth, improvements in standards of living and the economic growth that is deemed necessary (especially in LEDCs and NICs), it is suggested that efficiency would have to be raised by a factor of 4–10 to remain within global carrying capacity.

Many forms of sustainable development, such as sustainable urban development, are very expensive. While it is still cheaper to extract fossil fuels than develop alternative technologies (e.g. hydrogen fuel), governments and oil companies will continue to take the easy options. Governments face many pressures and calls on their resources – investing in technologies that might not be paid back for many decades is not seen as a vote winner.

■ QUICK CHECK QUESTIONS

26 Explain the meaning of the phrase *globalised consumer culture*.

27 Briefly explain why increasing standards of living lead to greater resource consumption.

EXAM PRACTICE

9 Explain why some people believe that the EFs of some countries need to be reduced. Justify whether an ecocentric or a technocentric approach to reducing the EF is more likely to be successful. [9]

10 Discuss how development policies and cultural influences can affect human population dynamics and growth. [8]

Common mistake

Reducing population growth does not necessarily lead to lower resource consumption – usually it coincides with increased consumption.

Expert tip

As countries develop, their ecological footprint generally increases, and the need for sustainable forms of development increases. The world could support many more people living at the EF of the 'average' Indian or Afghan citizen, compared with the 'average' American citizen.

Expert tip

You should be able to:

* **evaluate** the application of carrying capacity to local and global human populations
* **compare and contrast** the differences in the EFs of two countries
* **evaluate** how EVSs impact the EFs of individuals or populations.

Are you ready?

Use this checklist to record progress as you revise. Tick each box when you have:

- revised and understood a topic
- tested yourself using the **Quick check questions**
- used the **Exam practice** questions and gone online to check your answers.

	Revised	Tested	Exam ready
Topic 1 Foundations of environmental systems and societies			
1.1 Environmental value systems			
Development of the modern environmental movement	☐	☐	☐
Social systems	☐	☐	☐
Range of environmental value systems	☐	☐	☐
Decision making on environmental issues	☐	☐	☐
Environmental value systems of different societies	☐	☐	☐
Personal viewpoints on environmental issues	☐	☐	☐
Intrinsic value	☐	☐	☐
1.2 Systems and models			
Concept and characteristics of systems	☐	☐	☐
Transfer and transformation processes	☐	☐	☐
The systems concept on a range of scales	☐	☐	☐
Open systems, closed systems and isolated systems	☐	☐	☐
Models	☐	☐	☐
1.3 Energy and equilibria			
The first and second laws of thermodynamics and their relevance to environmental systems	☐	☐	☐
The nature of equilibria	☐	☐	☐
Positive feedback and negative feedback	☐	☐	☐
Tipping points	☐	☐	☐
Resilience of a system	☐	☐	☐

	Revised	Tested	Exam ready
1.4 Sustainability			
Resources and natural income	☐	☐	☐
Sustainability, natural capital and natural income	☐	☐	☐
Sustainable development	☐	☐	☐
Millennium Ecosystem Assessment (MA)	☐	☐	☐
Environmental impact assessments	☐	☐	☐
Ecological footprints	☐	☐	☐
1.5 Humans and pollution			
The nature of pollution	☐	☐	☐
The major sources of pollutants	☐	☐	☐
Pollution management: the process of pollution and strategies for reducing impacts	☐	☐	☐
Human factors that affect the approaches to pollution management	☐	☐	☐
The costs and benefits to society of the WHO's ban on the use of DDT	☐	☐	☐
Topic 2 Ecosystems and ecology			
2.1 Species and populations			
Species, populations, habitats and niches	☐	☐	☐
Abiotic and biotic factors	☐	☐	☐
Population interactions	☐	☐	☐
Limiting factors and carrying capacity	☐	☐	☐
S- and J-population curves	☐	☐	☐
2.2 Communities and ecosystems			
Communities and ecosystems	☐	☐	☐
Photosynthesis and respiration	☐	☐	☐
Food chains and food webs	☐	☐	☐

	Revised	Tested	Exam ready
Pyramids of numbers, biomass and productivity	☐	☐	☐
Pyramid structure and ecosystem functioning	☐	☐	☐
2.3 Flows of energy and matter			
Pathway of energy entering the atmosphere	☐	☐	☐
Transfer and transformation of energy	☐	☐	☐
Gross productivity, net productivity, primary productivity and secondary productivity	☐	☐	☐
Gross primary productivity and net primary productivity	☐	☐	☐
Gross secondary productivity and net secondary productivity	☐	☐	☐
Maximum sustainable yield	☐	☐	☐
Transfer and transformation of materials within an ecosystem	☐	☐	☐
Human impacts on energy flows and matter cycles	☐	☐	☐
2.4 Biomes, zonation and succession			
Biomes	☐	☐	☐
The tricellular model of atmospheric circulation	☐	☐	☐
The effect of climate change on biomes	☐	☐	☐
Succession	☐	☐	☐
Density-dependent and density-independent factors	☐	☐	☐
Changes through a succession	☐	☐	☐
Climax communities	☐	☐	☐
Ecosystem stability and succession	☐	☐	☐
Zonation	☐	☐	☐
2.5 Investigating ecosystems			
Measuring abiotic components of the system	☐	☐	☐
Identifying organisms in ecosystems	☐	☐	☐
Measuring biotic components of the system	☐	☐	☐

	Revised	Tested	Exam ready
Methods for estimating the abundance of organisms	☐	☐	☐
Method for estimating the biomass of trophic levels	☐	☐	☐
Diversity and the Simpson's diversity index	☐	☐	☐
Measuring changes along an environmental gradient	☐	☐	☐
Measuring changes due to a specific human activity	☐	☐	☐

Topic 3 Biodiversity and conservation

3.1 An introduction to biodiversity

Biodiversity	☐	☐	☐
Diversity indices	☐	☐	☐

3.2 Origins of biodiversity

Evolution	☐	☐	☐
The role of isolation in forming new species	☐	☐	☐
Plate activity	☐	☐	☐
Past and present rates of species extinction	☐	☐	☐

3.3 Threats to biodiversity

The number of species on Earth	☐	☐	☐
Factors that lead to loss of diversity	☐	☐	☐
Tropical biomes and sustainable development	☐	☐	☐
Factors that make species prone to extinction	☐	☐	☐
Extinct, critically endangered and back from the brink	☐	☐	☐
Natural area of biological significance under threat	☐	☐	☐

3.4 Conservation of biodiversity

Arguments for preserving species and habitats	☐	☐	☐
The role of intergovernmental and non-governmental organisations	☐	☐	☐
International conventions on biodiversity	☐	☐	☐
Designing a protected area	☐	☐	☐

	Revised	Tested	Exam ready

	Revised	Tested	Exam ready
Strengths and weaknesses of the species-based approach to conservation	☐	☐	☐
A mixed approach to conservation	☐	☐	☐
Topic 4 Water, aquatic food production systems and societies			
4.1 Introduction to water systems			
The hydrological cycle	☐	☐	☐
Ocean circulation systems	☐	☐	☐
4.2 Access to freshwater			
Access to freshwater	☐	☐	☐
4.3 Aquatic food production systems			
Demand for aquatic food resources	☐	☐	☐
Unsustainable and sustainable fishing practices	☐	☐	☐
4.4 Water pollution			
Sources of freshwater and marine pollution	☐	☐	☐
Biodegradation of organic material	☐	☐	☐
Indicator species	☐	☐	☐
Biotic indices	☐	☐	☐
Eutrophication	☐	☐	☐
Pollution management strategies	☐	☐	☐
Topic 5 Soil systems and terrestrial food production systems and societies			
5.1 Introduction to soil systems			
The soil system	☐	☐	☐
Soil structure and texture	☐	☐	☐

	Revised	Tested	Exam ready
5.2 Terrestrial food production systems and food choices			
Sustainability	☐	☐	☐
Inequalities	☐	☐	☐
Food distribution	☐	☐	☐
Factors influencing food production systems	☐	☐	☐
Food yield from different trophic layers	☐	☐	☐
Cultural choices	☐	☐	☐
Terrestrial food production systems	☐	☐	☐
Contrasting food production systems	☐	☐	☐
Links between social systems and food production	☐	☐	☐
Food sustainability	☐	☐	☐
5.3 Soil degradation and conservation			
Soil fertility and succession	☐	☐	☐
Soil degradation	☐	☐	☐
Soil conservation measures	☐	☐	☐
Topic 6 Atmospheric systems and societies			
6.1 Introduction to the atmosphere			
The atmosphere is a dynamic system	☐	☐	☐
The greenhouse effect	☐	☐	☐
6.2 Stratospheric ozone			
The role of ozone	☐	☐	☐
Ozone-depleting substances (ODSs)	☐	☐	☐
6.3 Photochemical smog			
Primary and secondary pollutants	☐	☐	☐
Pollution management strategies	☐	☐	☐

	Revised	Tested	Exam ready
6.4 Acid deposition			
The causes of acid deposition	☐	☐	☐
The effects of acid deposition	☐	☐	☐
Pollution management strategies	☐	☐	☐
Topic 7 Climate change and energy production			
7.1 Energy choices and security			
Energy sources	☐	☐	☐
The advantages and disadvantages of contrasting energy sources	☐	☐	☐
Energy security	☐	☐	☐
Energy choices	☐	☐	☐
Energy efficiency and energy conservation	☐	☐	☐
7.2 Climate change – causes and impacts			
Climate and weather	☐	☐	☐
Atmospheric and oceanic circulatory systems	☐	☐	☐
Greenhouse gases and the greenhouse effect	☐	☐	☐
Human activities and greenhouse gases	☐	☐	☐
The impacts of climate change	☐	☐	☐
Positive and negative feedbacks	☐	☐	☐
Complexity of climate change	☐	☐	☐
Global dimming	☐	☐	☐
7.3 Climate change – mitigation and adaptation			
Mitigation strategies	☐	☐	☐
Adaptation strategies	☐	☐	☐
International agreements	☐	☐	☐
Individuals' reductions in greenhouse gas emissions	☐	☐	☐

	Revised	Tested	Exam ready
Topic 8 Human systems and resource use			
8.1 Human population dynamics			
Demographic indicators	☐	☐	☐
Global population change	☐	☐	☐
Population growth and resources	☐	☐	☐
Age–sex (population) pyramids	☐	☐	☐
The demographic transition model	☐	☐	☐
Factors affecting population dynamics	☐	☐	☐
8.2 Resource use in society			
Renewable and non-renewable natural capital	☐	☐	☐
Natural capital – goods and services	☐	☐	☐
The dynamic nature of natural capital	☐	☐	☐
8.3 Solid domestic waste			
Types of solid domestic waste	☐	☐	☐
Non-biodegradable waste	☐	☐	☐
Waste disposal methods	☐	☐	☐
Pollution management strategies	☐	☐	☐
8.4 Human population carrying capacity			
Carrying capacity	☐	☐	☐
Ecological footprints	☐	☐	☐
Degradation, consumption and population growth	☐	☐	☐
The role of technology	☐	☐	☐

Glossary

Abiotic factor A non-living, physical factor that can influence an organism or ecosystem – for example, temperature, sunlight, pH, salinity or precipitation.

Acute pollution Pollution that produces its effects through a short, intense exposure. Symptoms are usually experienced within hours.

Adaptive radiation Where an ancestral species evolves to fill different ecological niches, leading to new species.

Age–sex pyramid A bar graph that shows the relative or absolute amount of people in a population at different ages and of different sexes.

Albedo A measure of the reflecting power of a surface in relation to the amount of short-wave radiation received. The amount of light reflected by a surface.

Aquifer Water-bearing rock.

Bacteria Very small, single-celled organisms that do not have a true nucleus.

Bioaccumulation The build-up of persistent or non-biodegradable pollutants within an organism or trophic level because they cannot be broken down.

Biochemical oxygen demand (BOD) A measure of the amount of dissolved oxygen required to break down the organic material in a given volume of water through aerobic biological activity.

Biodegradable Capable of being broken down by natural biological processes – for example, through the activities of decomposer organisms.

Biodiversity The amount of biological or living diversity per unit area. It includes the concepts of species diversity, habitat diversity and genetic diversity.

Biomagnification The increase in concentration of persistent or non-biodegradable pollutants along a food chain.

Biomass The mass of organic material in organisms or ecosystems, usually stated per unit area.

Biome A collection of ecosystems sharing similar climatic conditions – for example, tundra, tropical rainforest and desert.

Biosphere That part of the Earth inhabited by organisms.

Biotic factor A living part of an ecosystem (i.e. part of the community) that can influence an organism or ecosystem.

Carrying capacity The maximum number of a species or 'load' that can be sustainably supported by a given environment.

Chronic pollution Pollution that produces its effects through low-level, long-term exposure. Disease symptoms develop up to several decades later.

Climax community A community of organisms that is more or less stable, and that is in equilibrium with natural environmental conditions such as climate. It is the end point of ecological succession.

Closed system A system that exchanges only energy but not matter with its surroundings (e.g. the Earth).

Community A group of populations living and interacting with each other in a common habitat.

Competition A common demand by two or more organisms for a limited supply of a resource such as food, water, light, space, mates and nesting sites.

Continental Relating to areas that are distant from the coast.

Crude birth rate (CBR) The number of live births per 1000 people in a population per year.

Crude death rate (CDR) The number of deaths per 1000 people in a population per year.

Demographic transition model A model that shows the change in a population from one that has high birth rates and high death rates to a country that has low birth rates and low death rates.

Desertification The spread of desert-like conditions into previously green areas, causing a long-term decline in biological productivity.

Diversity A generic term for heterogeneity (i.e. variation or variety). The scientific meaning of diversity becomes clear from the context in which it is used; it can refer to heterogeneity of species or habitat, or to genetic heterogeneity.

Doubling time The length of time it takes for a population to double in size, assuming its natural growth rate remains constant. Approximate values for it can be obtained by using the formula:

$$\text{doubling time (years)} = \frac{70}{\text{growth rate (\%)}}$$

Ecological footprint (EF) The area of land and water required to support a defined human population at a given standard of living. The measure takes account of the area required to provide all the resources needed by the population, and the assimilation of wastes.

Ecological value Resources with no formal market price: soil erosion control, nitrogen fixation and photosynthesis are all essential for human existence but have no direct monetary value, although some estimates have been made.

Economic value The market price of the goods and services a resource produces.

Economic water scarcity Where water is available locally, but not accessible for human, institutional or financial capital reasons.

Ecosystem A community and the physical environment with which it interacts.

Eluviation The removal of material down a soil through solution and suspension.

Energy security Having an adequate, reliable and affordable supply of energy.

Entropy A measure of the amount of disorder, chaos or randomness in a system; the greater the disorder, the higher the level of entropy.

Environmental impact assessment (EIA) A method of detailed survey required, in many countries, before a major development. Ideally it should be independent of, but paid for by, the developer. It should include a baseline study, and monitoring should continue after completion of the project.

Environmental impact statement Report produced by an **environmental impact assessment**.

Environmental value system A particular worldview that shapes the way an individual or group of people perceives and evaluates environmental issues, influenced by cultural, religious, economic and sociopolitical contexts.

Equilibrium A state of balance among the components of a system.

Eutrophication The natural or artificial enrichment of a body of water, particularly with respect to nitrates and phosphates, that results in depletion of oxygen levels in the water. Eutrophication is accelerated by human activities that add detergents, sewage or agricultural fertilisers to bodies of water.

Evolution The cumulative, gradual change in the genetic characteristics of successive generations of a species or race of an organism, ultimately giving rise to species or races different from the common ancestor. Evolution reflects changes in the genetic composition of a population over time.

Exponential growth An increasing or accelerating rate of growth, sometimes referred to as a J-shaped or J-population curve.

Extrapolation technique Estimating values beyond measured values, using graphs or other techniques.

Feedback When part of the output from a system returns as an input, so as to affect subsequent outputs.

Fossil fuels Non-renewable resources including oil, coal, natural gas and shale gas.

Fundamental niche The full range of conditions and resources in which a species could survive and reproduce.

Fungi Organisms that are heterotrophic and have cell walls made of chitin.

Genetic diversity The range of genetic material present in a population of a species.

Global warming An increase in average temperature of the Earth's atmosphere.

Greenhouse gases Atmospheric gases that absorb infrared radiation, causing world temperatures to be warmer than they would otherwise be. This process is sometimes known as 'radiation trapping'. The natural greenhouse effect is caused mainly by water and carbon dioxide.

Gross primary productivity (GPP) The total gain in energy or biomass per unit area per unit time fixed by photosynthesis in green plants.

Gross productivity (GP) The total gain in energy or biomass per unit area per unit time, which could be through photosynthesis in primary producers or absorption in consumers.

Gross secondary productivity (GSP) The total gain by consumers in energy or biomass per unit area per unit time through absorption.

Groundwater Water that is found beneath the Earth's surface in water-bearing rocks.

Habitat The environment in which a species normally lives.

Habitat diversity The range of different habitats in an ecosystem or biome. Conservation of habitat diversity usually leads to the conservation of species and genetic diversity.

Halogenated organic gases Usually known as halocarbons, these were first identified as depleting the ozone layer in the stratosphere. They are now known to be potent greenhouse gases. The best-known are the chlorofluorocarbons (CFCs).

Humus Partially decomposed organic matter derived from the decay of dead plants and animals in soils.

Hydrological cycle The cycle of water between the atmosphere, lithosphere and biosphere.

Illuviation The redeposition of material in the lower horizons.

Indicator species A species whose presence, absence or abundance can be used as an indication of pollution.

Infant mortality rate (IMR) The number of deaths of children less than 1 year old per 1000 live births per year.

Intrinsic value A characteristic of a natural system that has an inherent worth, irrespective of economic considerations, such as the belief that all life on Earth has a right to exist.

Invasive species Introduction of non-native species.

Isolated system A system that does not exchange either matter or energy with its surroundings (e.g. the Universe).

Isolation The process by which two populations become separated by geographical, behavioural, genetic or reproductive factors. If gene flow between the two subpopulations is prevented, new species may evolve. See also **evolution**.

J-population curve Population growth curve which shows exponential growth. Growth is initially slow, then increasingly rapid, and does not slow down.

K-strategists Species that usually concentrate their reproductive investment in a small number of offspring, thus increasing their survival rate and adapting them for living in long-term climax communities.

Latitude The angular distance from the equator (north or south of it) as measured from the centre of the Earth (usually in degrees).

Leaching The removal of soluble material in solution.

Life expectancy (E0) The average number of years that a person can be expected to live, usually from birth, if demographic factors remain unchanged.

Maritime Relating to areas that are close to the coast.

Mass extinction Events in which 75% of the species on Earth disappear within a geologically short time period, usually between a few hundred thousand to a few million years.

Maximum sustainable yield The rate of increase in natural capital, i.e. that which can be exploited without depleting the original stock or its potential from replenishment.

Model A simplified version of a system. It shows the flows and storages as well as the structure and workings.

Motile organism One that can actively move under its own power from place to place.

Natural capital Natural resources that are managed to provide a sustainable natural income from goods or services.

Natural income The portion of natural capital (resources) that is produced as 'interest', i.e. the sustainable income produced by natural capital.

Natural increase (annual growth rate) Found by subtracting the crude death rate (‰ – per thousand) from the crude birth rate (‰) and is then expressed as a percentage (%):

Negative feedback Feedback that tends to counteract any deviation from equilibrium and promotes stability.

Net primary productivity (NPP) The gain by producers in energy or biomass per unit area per unit time remaining after allowing for respiratory losses (R). This is potentially available to consumers in an ecosystem.

Net productivity (NP) The gain in energy or biomass per unit area per unit time remaining after allowing for respiratory losses (R).

Net secondary productivity (NSP) The gain by consumers in energy or biomass per unit area per unit time remaining after allowing for respiratory losses (R).

Niche A species' share of a habitat and the resources in it. An organism's ecological niche depends not only on where it lives but also on what it does.

Non-motile organism One that cannot move or, for the purposes of sampling, can only move very slowly (such as limpets on the rocky shore).

Non-point-source pollution The release of pollutants from numerous, widely dispersed origins – for example, gases from the exhaust systems of vehicles or power plants.

Non-renewable Natural resources that cannot be replenished within a timescale of the same order as that at which they are taken from the environment and used – for example, fossil fuels.

Ocean conveyor belt The deep, large-scale circulation of the ocean's waters that is largely responsible for the transfer of heat from the tropics to colder regions.

Omnivore An organism that eats both plants and animals.

Open system A system that exchanges both matter and energy with its surroundings (e.g. an ecosystem).

Persistent organic pollutants (POPs) Chemical substances that persist in the environment, bioaccumulate through the food web, and pose a risk of causing adverse effects to human health and the environment.

Physical water scarcity Where water resource development is approaching, or has exceeded, unsustainable levels. It relates water availability to water demand and implies that arid areas are not necessarily water scarce.

Pioneer community The first stage of an ecological succession that contains hardy species able to withstand difficult conditions.

Plate tectonics The movement of the eight major and several minor internally rigid plates of the Earth's lithosphere in relation to each other and to the partially mobile **asthenosphere** below.

Point-source pollution The release of pollutants from a single, clearly identifiable site – for example, a factory chimney or the waste disposal pipe of a factory into a river.

Pollution The addition of a substance or an agent to an environment through human activity, at a rate greater than that at which it can be rendered harmless by the environment, and which has an appreciable effect on the organisms in the environment.

Population A group of organisms of the same species living in the same area at the same time, and which are capable of interbreeding.

Positive feedback Feedback that increases change; it promotes deviation away from an equilibrium.

Primary pollutant A pollutant that is active on emission.

Primary productivity The gain by producers in energy or biomass per unit area per unit time. This term could refer to either gross or net primary productivity.

Pro-natalist population policy A policy that is in favour of more births.

Realised niche The actual conditions and resources in which a species exists due to biotic interactions.

Regolith The irregular cover of loose rock debris that covers the Earth.

Renewable Natural resources that have a sustainable yield or harvest equal to, or less than, their natural productivity – for example, timber.

Resilience The tendency of a system to avoid tipping points and maintain stability through steady-state equilibrium

r-strategists Species that tend to spread their reproductive investment among a large number of offspring so that they are well adapted to colonise new habitats rapidly and make opportunistic use of short-lived resources.

Secondary pollutant A pollutant that arises from a primary pollutant that has undergone physical or chemical change.

Secondary productivity The biomass gained by consumers (heterotrophic organisms), through feeding and absorption, measured in units of mass or energy per unit area per unit time.

Sere The set of communities that succeeds another over the course of succession at a given location.

Smog The term now used for any haziness in the atmosphere caused by air pollutants. Photochemical smog is produced through the effect of ultraviolet light on the products of internal combustion engines. It may contain ozone and is damaging to the human respiratory system and eyes.

Social system People, groups and institutions that work together, forming distinct patterns and relationships that define the society.

Society An arbitrary group of individuals who share some common characteristic such as geographical location, cultural background, historical timeframe, religious perspective, value system and so on.

Soil A mixture of mineral particles and organic material that covers the land, and in which terrestrial plants grow.

Soil profile A vertical section through a soil, from the surface down to the parent material, revealing the soil layers or horizons.

Speciation The formation of new species when populations of a species become isolated and evolve differently from other populations. See also **evolution**.

Species A group of organisms that interbreed and are capable of producing fertile offspring.

Species diversity The variety of species per unit area. This includes both the number of species present and their relative abundance.

Stable equilibrium The tendency in a system for it to return to a previous equilibrium condition following disturbance. This is in contrast to **unstable equilibrium**, which forms a new equilibrium following disturbance.

Steady-state equilibrium The condition of an open system in which there are no changes over the longer term, but in which there may be oscillations in the very short term.

Succession The orderly process of change *over time* in a community.

Sustainability The use of global resources at a rate that allows natural regeneration and minimises damage to the environment.

Sustainable development Development that meets current needs without compromising the ability of future generations to meet their own needs.

Sustainable yield When a natural resource can be harvested at a rate equal to or less than its natural productivity so that the natural capital is not diminished.

System An assemblage of parts and the relationships between them, which together constitute an entity or whole.

Tipping point A critical threshold when even a small change can have dramatic effects and cause a disproportionately large response in the overall system.

Total fertility rate (TFR) The average number of births per woman of childbearing age.

Trophic level The position that an organism occupies in a food chain, or a group of organisms in a community that occupy the same position in food chains.

Virtual water Water that is used to produce crops or flowers, and then the product is exported (often from LEDCs to MEDCs).

Zonation The arrangement or patterning of plant communities or ecosystems into parallel or sub-parallel bands in response to change, over a distance, in some environmental factor.